Ready, Set, Go!™

Numbers & Operations

Workbook

Mel Friedman, M.S.

Research & Education Association
Visit our website at
www.rea.com

Research & Education Association

61 Ethel Road West
Piscataway, New Jersey 08854
E-mail: info@rea.com

REA's Ready, Set, Go!™
Numbers and Operations Workbook

Printed in the United States of America

ISBN-13: 978-0-7386-0451-0
ISBN-10: 0-7386-0451-8

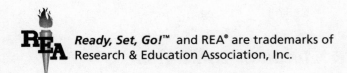

Ready, Set, Go!™ and REA® are trademarks of
Research & Education Association, Inc.

Contents

INTRODUCTION

About This Book...vii

Icons Explained..vii

Letters to Students, Parents, and Teachers...............................viii

About REA x

Acknowledgments x

LESSON 1

Place Value for Whole Numbers and Decimals..........................1

Place value for whole numbers 2

Place value for decimals 3

Test Yourself! 6

LESSON 2

Rounding Off Whole Numbers 8

Test Yourself! 12

LESSON 3

Rounding Off Decimal Numbers...13

Test Yourself! 17

LESSON 4

Adding/Subtracting with Signed Numbers 19

Test Yourself! 23

LESSON 5

Multiplying/Dividing with Signed Numbers...........................24

Test Yourself! 27

QUIZ ONE (Lessons 1–5) 29

LESSON 6

Reducing Fractions to Lowest Terms 31

Test Yourself! 36

LESSON 7

Converting Improper Fractions to Mixed Fractions and Vice Versa...38

Changing an improper fraction to a mixed fraction 39

Changing a mixed fraction to an improper fraction 40

Test Yourself! 43

LESSON 8

Adding /Subtracting Fractions...**44**

Addition and subtraction with common denominators................................45

Addition and subtraction with different denominators 45

Test Yourself! 50

LESSON 9

Multiplying/Dividing Fractions..**52**

Multiplication of fractions...53

Division of fractions...56

Test Yourself! 58

QUIZ TWO (Lessons 6–9) **59**

LESSON 10

Converting Percents to Fractions and Decimals **61**

Test Yourself! 66

LESSON 11

Converting Decimals to Fractions and Percents **67**

Test Yourself! 72

LESSON 12

Converting Fractions to Decimals and Percents **73**

Test Yourself! 77

LESSON 13

Percent Increase and Decrease **78**

Percent increase 79

Percent decrease 80

Test Yourself! 84

LESSON 14

Comparing Sizes of Numbers in Decimal/Fraction Form.........................**86**

Comparing sets of decimals...87

Comparing sets of fractions 87

Test Yourself! 90

QUIZ THREE (Lessons 10–14) **91**

LESSON 15

Exponents...**93**

Test Yourself! 96

LESSON 16

Scientific Notation.. 98
Changing numbers to scientific notation................................. 99
Changing scientific notation to an integer or decimal 101
Test Yourself! 102

LESSON 17

Order of Operations... 104
Examples with no parentheses.. 105
Examples with operations in parentheses................................. 106
Test Yourself! 108

QUIZ FOUR (Lessons 15–17) 109

LESSON 18

Divisibility Rules for Integers.. 111
Rules for 1 to 4 112
Rules for 5 to 8 113
Rules for 9 and 10.. 114
Test Yourself! 114

LESSON 19

Primes and Composites... 117
Definitions of prime and composite numbers........................... 118
Factors of numbers 118
Test Yourself! 120

LESSON 20

Prime Factorization 122
Definition of prime factorization.. 123
Writing the prime factorization for a given number 123
Test Yourself! 125

LESSON 21

Greatest Common Factor (GCF) 127
Definition of greatest common factor 128
Determining the greatest common factor 128
Test Yourself! 131

LESSON 22

Least Common Multiple (LCM) 132
Definition of least common multiple 133
Determining the least common multiple 133
Test Yourself! 135

LESSON 23
  The Relationship between Numbers and Letters....................................137
  Translation of word phrases into numerical symbols 138
  Test Yourself! 141

QUIZ FIVE (Lessons 18–23) 143

CUMULATIVE EXAM....................................145

ANSWER KEY 149

SCORECARD 182

Welcome
to the *Ready, Set, Go!*
Numbers & Operations Workbook!

About This Book

This book will help high school math students at all learning levels understand basic mathematics. Students will develop the skills, confidence, and knowledge they need to succeed on high school math exams with emphasis on passing high school graduation exams.

More than 20 easy-to-follow lessons break down the material into the basics. In-depth, step-by-step examples and solutions reinforce student learning, while the "Math Flash" feature provides useful tips and strategies, including advice on common mistakes to avoid.

Students can take drills and quizzes to test themselves on the subject matter, then review any areas in which they need improvement or additional reinforcement. The book concludes with a final exam, designed to comprehensively test what students have learned.

The *Ready, Set, Go! Numbers & Operations Workbook* will help students master the basics of mathematics—and help them face their next math test—with confidence!

Icons Explained

Icons make navigating through the book easier. The icons, explained below, highlight tips and strategies, where to review a topic, and the drills found at the end of each lesson.

 Look for the **"Math Flash"** feature for helpful tips and strategies, including advice on how to avoid common mistakes.

 When you see the **"Let's Review"** icon, you know just where to look for more help with the topic on which you are currently working.

 The **"Test Yourself!"** icon, found at the end of every lesson, signals a short drill that reviews the skills you have studied in that lesson.

To the Student

This workbook will help you master the fundamentals of Numbers & Operations. It offers you the support you need to boost your skills and helps you succeed in school and beyond!

It takes the guesswork out of math by explaining what you most need to know in a step-by-step format. When you apply what you learn from this workbook, you can

1. do better in class;

2. raise your grades, and

3. score higher on your high school math exams.

Each compact lesson in this book introduces a math concept and explains the method behind it in plain language. This is followed with lots of examples with fully worked-out solutions that take you through the key points of each problem.

The book gives you two tools to measure what you learn along the way:

✔ **Short drills that follow <u>each</u> lesson**

✔ **Quizzes that test you on <u>multiple</u> lessons**

These tools are designed to comfortably build your test-taking confidence.

Meanwhile, the "Math Flash" feature throughout the book offers helpful tips and strategies—including advice on how to avoid common mistakes.

When you complete the lessons, take the final exam at the end of the workbook to see how far you've come. If you still need to strengthen your grasp on any concept, you can always go back to the related lesson and review at your own pace.

To the Parent

For many students, math can be a challenge—but with the right tools and support, your child can master the basics of mathematics. As educational publishers, our goal is to help all students develop the crucial math skills they'll need in school and beyond.

This *Ready, Set, Go! Workbook* is intended for students who need to build their basic mathematics skills. It was specifically created and designed to assist students who need a boost in understanding and learning the math concepts that are most tested along the path to graduation. Through a series of easy-to-follow lessons, students are introduced to the essential mathematical ideas and methods, and then take short quizzes to test what they are learning.

Each lesson is devoted to a key mathematical building block. The concepts and methods are fully explained, then reinforced with examples and detailed solutions. Your child will be able to test what he or she has learned along the way, and then take a cumulative exam found at the end of the book.

Whether used in school with teachers, for home study, or with a tutor, the ***Ready, Set, Go! Workbook*** is a great support tool. It can help improve your child's math proficiency in a way that's fun and educational!

To the Teacher

As you know, not all students learn the same, or at the same pace. And most students require additional instruction, guidance, and support in order to do well academically.

Using the Curriculum Focal Points of the National Council of Teachers of Mathematics, this workbook was created to help students increase their math abilities and succeed on high school exams with special emphasis on high school proficiency exams. The book's easy-to-follow lessons offer a review of the basic material, supported by examples and detailed solutions that illustrate and reinforce what the students have learned.

To accommodate different pacing for students, we provide drills and quizzes throughout the book to enable students to mark their progress. This approach allows for the mastery of smaller chunks of material and provides a greater opportunity to build mathematical competence and confidence.

When we field-tested this series in the classroom, we made every effort to ensure that the book would accommodate the common need to build basic math skills as effectively and flexibly as possible. Therefore, this book can be used in conjunction with lesson plans, stand alone as a single teaching source, or be used in a group-learning environment. The practice quizzes and drills can be given in the classroom as part of the overall curriculum or used for independent study. A cumulative exam at the end of the workbook helps students (and their instructors) gauge their mastery of the subject matter.

We are confident that this workbook will help your students develop the necessary skills and build the confidence they need to succeed on high school math exams.

About Research & Education Association

Founded in 1959, Research & Education Association (REA) is dedicated to publishing the finest and most effective educational materials—including software, study guides, and test preps—for students in elementary school, middle school, high school, college, graduate school, and beyond.

Today, REA's wide-ranging catalog is a leading resource for teachers, students, and professionals.

We invite you to visit us at *www.rea.com* to find out how "REA is making the world smarter."

About the Author

Author Mel Friedman is a former classroom teacher and test-item writer for Educational Testing Service and ACT, Inc.

Acknowledgments

We would like to thank Larry Kling, Vice President, Editorial, for his editorial direction; Pam Weston, Vice President, Publishing, for setting the quality standards for production integrity and managing the publication to completion; Alice Leonard, Senior Editor, for project management and preflight editorial review; Diane Goldschmidt, Senior Editor, for post-production quality assurance; Ruth O'Toole, Production Editor, for proofreading; Rachel DiMatteo, Graphic Artist, for page design; Christine Saul, Senior Graphic Artist, for cover design; and Jeff LoBalbo, Senior Graphic Artist, for post-production file mapping.

We also gratefully acknowledge Heather Brashear for copyediting, and Kathy Caratozzolo of Caragraphics for typesetting.

Our sincere gratitude is extended to longtime math tutor, teacher, and author Bob Miller for his review of all test items.

A special thank you to Muriel Coletta, Bill Klimas, Sabat Upendra, Carolyn Mehlhorn, and the students of Plainfield High School in Plainfield, New Jersey, for reviewing and/or field-testing lessons from this book.

Place Value for Whole Numbers and Decimals

In this lesson, we will explore the meaning of the individual digits of a number, whether it is a whole number or a decimal. This skill is very valuable.

The ability to put numbers into words and vice versa is important to our daily lives. Think about what would happen if a bank accidentally dropped decimal points. A person's bank account of $12,400.05 might be recorded as $1,240,005—a difference of one million dollars! As another example, suppose you pay $2.90 for a gallon of gasoline. Imagine your shock if the price read $290 instead. As a third example, suppose a surgeon was to perform an operation in which he must make a cut of 0.03 inch in a person's body. If the surgeon mistakenly made a 0.3 inch cut, the result could be serious.

Your Goal: When you have completed this lesson, you should be able to identify the meaning and interpretation of individual digits of any number.

LESSON 1

Place Value for Whole Numbers and Decimals

For whole numbers, reading from right to left, the names of the place values are as follows:

Billions	Hundred-Millions	Ten-Millions	Millions	Hundred-Thousands	Ten-Thousands	Thousands	Hundreds	Tens	Ones
1,	2	3	4,	5	6	7,	8	9	0

This list is endless, but you will rarely see numbers with higher values.

1

Example: *How should the number 507 be written in words?*

Solution: The "5" represents "hundreds," the "0" represents "tens," and the "7" represents "ones." This number is written as "five hundred seven." (It could be written as "five hundred and seven." This is correct, but it is not necessary to use the word "and.")

2

Example: *How should the number 2148 be written in words?*

Solution: The "2" represents "thousands," the "1" represents "hundreds," the "4" represents "tens," and the "8" represents "ones." This number is written as "two thousand one hundred forty-eight."

3

Example: *How should the number 325,089 be written in words?*

Solution: Instead of identifying each digit separately, the key is to realize that the highest place value, which is the digit 3, is in the hundred-thousands place. Thus, this number is written as "three hundred twenty-five thousand eighty-nine."

Any number with a value of at least 10,000 should be written with commas that separate each group of three digits, counting from the right. Therefore, 35078162 would be written as 35,078,162.

4

Example: *How should the number 41,590 be written in words?*

Solution: Since this number has five digits, the digit 4 of this number represents ten-thousands. Combine the 4 and 1 and write it as "forty-one thousand." Thus, we write "forty-one thousand five hundred ninety."

A **decimal number** has at least one non-zero digit to the right of the decimal point. The numbers 0.6, 5.407, and 34.1179 are examples.

For decimal numbers, reading from left to right from the placement of the decimal point, the place values are as follows:

Ones	Tenths	Hundredths	Thousandths	Ten-Thousandths	Hundred-Thousandths	Millionths
0 .	1	2	3	4	5	6

This list is also endless, but you will hardly ever see units of numbers smaller than millionths. If there is no digit to the left of the decimal point, a "0" is placed in that spot. The use of "0" helps the reader to easily recognize the number as a decimal.

5

Example: *How should the number 0.49 be written in words?*

Solution: The "4" represents "tenths," and the "9" represents "hundredths." This number is written as "forty-nine hundredths." Notice that "ths" at the end of "hundredths" signals that positions to the right of the decimal point are occupied by digits (in the case of hundredths, two positions are occupied).

6

Example: *How should the number 0.0651 be written in words?*

Solution: The first "0" to the right of the decimal point represents "tenths," the "6" represents "hundredths," the "5" represents "thousandths," and the "1" represents "ten-thousandths." The number is written as "six hundred fifty-one ten-thousandths."

7

Example: *How should the number 0.007 be written in words?*

Solution: Since there are three placeholders to the right of the decimal point, the answer must contain the word "thousandths." The number is written as "seven thousandths."

8

Example: *How should the number 0.29734 be written in words?*

Solution: Instead of identifying each digit separately, note that the right-most digit, which is 4, is in the hundred-thousandths place. The number is written as "twenty-nine thousand seven hundred thirty-four hundred-thousandths."

MathFlash!

Rarely will a decimal number have a zero as its rightmost digit. An example would involve rounding off a number, which will be discussed in a future lesson. If such a decimal appears, the position occupied by zero must be stated when writing the words for the number.

Example: *0.180 would be written as "one hundred eighty thousandths." Of course, the actual value of 0.180 is equivalent to 0.18 , which is written as "eighteen hundredths."*

Often, numbers are written in words and must be converted back to the corresponding digits. An example in which both words and digits need to be written would be in writing a check.

9

Example: *How should "three thousand sixteen" be written in digits?*

Solution: We need four place values to the left of the decimal point, beginning with 3. Since the word "sixteen" will occupy two place values, we will need a zero in the hundreds place. The answer is 3016.

10

Example: *How should "eight hundred eighty" be written in digits?*

Solution: Since the highest place value is hundreds, the number will have three digits. The answer is 880.

11

Example: *How should "forty-five thousand eight hundred thirty-two" be written in digits?*

Solution: The phrase "forty-five thousand" indicates that this number requires five place values, beginning with 4. "Forty-five" occupies two place values, and "eight hundred thirty-two" occupies three place values. The answer is 45,832.

12

Example: *How should the number "nine million three" be written in digits?*

Solution: "Nine million" would start in the seventh place value, but "three" only occupies one place value. Do you see why we will need five "fillers"? The answer is 9,000,003.

MathFlash!

Whole numbers may have many zeros, depending on where the leftmost digit is located. You will not need to put any zeros to the left of the first nonzero digit. For example, if the required answer is 783, there is no need to write 0783. Examples of circumstances in which "leading" zeros are necessary are identification labels and zip codes.

Example:

13

How should the number "fifty-one hundredths" be written in digits?

Solution: "Fifty-one" occupies two place values, and "hundredths" occupies two place values. Therefore, 0.51 is the answer.

Example:

14

How should the number "one hundred forty-seven ten-thousandths" be written in digits?

Solution: "One hundred forty-seven" will occupy three place values; however, "ten-thousandths" requires four place values. This means that a zero is needed immediately to the right of the decimal point. Then 0.0147 is the answer.

Example:

15

How should the number "five thousand three hundred twenty ten-thousandths" be written in digits?

Solution: As we noted following Example 6, there are instances in which a decimal may have a zero as its rightmost digit. The phrase "five thousand three hundred twenty" requires four place values, and "ten-thousandths" also requires four place values. Thus, 0.5320 is the answer.

Example:

16

The width of a very thin pencil point is sixty-five hundred-thousandths of a centimeter. How is this number written in digits?

Solution: A total of five digits to the right of the decimal point are needed. Since "sixty-five" occupies two place values, we will need three zeros. Then 0.00065 is the answer.

Test Yourself!

1.   Write the number 58,841 in words.

Answer: _____

2.   Write the number 9005 in words.

Answer: _____

Test Yourself! *(continued)*

3. Write the number 0.307 in words.

    Answer: _____

4. Write the number 0.6543 in words.

    Answer: _____

5. Write the number 0.00806 in words.

    Answer: _____

6. Write in digits "four hundred one thousand twenty-seven."

    Answer: _____

7. Write in digits "one thousand seven hundred fifteen."

    Answer: _____

8. Write in digits "eighty-one thousandths."

    Answer: _____

9. Write in digits "five hundred forty-two hundred-thousandths."

    Answer: _____

10. Write in digits "two thousand four hundred nine ten-thousandths."

    Answer: _____

Rounding Off Whole Numbers

In this lesson, we will explore how to round off whole numbers to a specified place value. You have read and heard about many different settings in which this procedure is done, such as (a) the number of fans at a soccer game, rounded off to the nearest thousand; (b) the weight of a bag of grapes, rounded off to the nearest pound; (c) the number of people living in New Jersey, rounded off to the nearest ten thousand; and (d) the time for a horse to run a mile, rounded off to the nearest second.

Your Goal: When you have completed this lesson, you should be able to approximate any number to a specified place value. In rounding off, you give up a small amount of accuracy to help make the numbers easier to work with.

LESSON 2

Rounding Off Whole Numbers

A **whole number** is zero or a counting number, such as 4, 17, 587, and so forth. The smallest counting number is 1. There are times when we need information on a number, but not an exact answer. For example, attendance at a baseball game is usually a number with five digits, such as 43,268. But, you may only need to know that the attendance was about 43,000. In this lesson, our objective will be to understand how to round off any counting number to a specified place.

Billions	Hundred-Millions	Ten-Millions	Millions	Hundred-Thousands	Ten-Thousands	Thousands	Hundreds	Tens	Ones
1,	2	3	4,	5	6	7,	8	9	0

Counting numbers are also called natural numbers.

Example: *What is 761 rounded off to the nearest ten?*

1

Solution: Remember that the tens digit is the second digit from the right of the decimal point. We understand that 761 means the number with a decimal point (761.). Automatically, the units digit, in this case 1, will change to 0. Since this units digit is less than 5, the digit in the tens place, which is 6, will stay the same. The final answer is 760.

2

Example: *What is 5483 rounded off to the nearest hundred?*

Solution: The hundreds digit is 4, so any digit to the right of 4 will automatically become 0. Now look at the digit in the tens place, which is 8. Since this digit is at least 5, the digit in the hundreds place, which is 4, will increase by one to become a 5. The final answer is 5500.

3

Example: *What is 7094 rounded off to the nearest thousand?*

Solution: The thousands digit is 7, so the three digits to its right all become zeros. Note that the hundreds digit is already zero. In fact, we need to look at the hundreds digit. Since this digit is less than 5, the digit in the thousands place stays the same. The final answer is 7000.

4

Example: *What is 17,992 rounded off to the nearest ten?*

Solution: The units digit automatically becomes 0, and since the units digit of 2 is less than 5, there is no change in the 9 that occupies the tens place. The final answer is 17,990.

5

Example: *What is 17,992 rounded off to the nearest hundred?*

Solution: The two digits that occupy the tens and units places, 9 and 2, must become zeros. The digit in the tens place is 9, so this will force the digit in the hundreds place to increase by one. However, notice that the digit in the hundreds place is also 9. When the digit 9 increases by one, it not only becomes 0, but it forces the digit in the thousands place, which is 7, to increase by one and become 8. The final answer is 18,000.

6

Example: *Let us return to the number 43,268 from the introduction, representing the attendance at a baseball game. What would this number be if it were rounded off to the nearest ten?*

Solution: The answer would be 43,270. To the nearest hundred, the answer would be 43,300. To the nearest thousand, you can see that the answer is 43,000.

7

Example: *What is 5483 rounded off to the nearest ten-thousand?*

Solution: Remember to first change every digit in a position that is lower than the ten-thousands place into zeros. However, notice that the digit in the thousands place is at least 5. This means that we must increase the hundred-thousands digit by one. Since there is no hundred-thousands digit, it is treated as if it were a zero. So, when it is increased by one, it becomes 1. The final answer is 10,000. The good news is that you will rarely be asked to round off to a position that exceeds the highest placeholder of the original number.

8

Example: *What is 12,451 rounded off to the nearest hundred?*

Solution: First, we will change both the tens digit and units digit to zeros. Next, the digit in the tens place (5) is at least 5, so the digit in the hundreds place must be increased by one. The final answer is 12,500.

9

Example: *What is 12,451 rounded off to the nearest thousand?*

Solution: Caution: Do NOT use the answer of Example 8 as a starting point! You must begin with the original number. Change the digits in the hundreds, tens, and units places to zeros. Now, the digit in the hundreds place, which is 4, is not at least 5. So, the digit in the thousands place (2) will not change. The final answer is 12,000. Notice the mistake you would have made if you simply used the result from Example 8; you would have incorrectly rounded off the answer to 13,000.

10

Example: *What is 283,950 rounded off to the nearest ten-thousand?*

Solution: The digits 3, 9, and 5 must change to zeros. (The units digit is already zero.) We check the digit in the thousands place, which is 3. Since this digit is less than 5, we do not change the 8. The final answer is 280,000.

MathFlash!

In any "rounding off" problem, be sure you first identify which digits automatically become zero. Then make sure you know which digit(s) will be affected by the rounding.

Test Yourself!

Round off each number to the specified place.

1.   **9107 to the nearest ten**   *Answer:* _____

2.   **2639 to the nearest hundred**   *Answer:* _____

3.   **49,661 to the nearest thousand**   *Answer:* _____

4.   **5403 to the nearest ten**   *Answer:* _____

5.   **748,492 to the nearest hundred**   *Answer:* _____

6.   **905,188 to the nearest thousand**   *Answer:* _____

7.   **396,222 to the nearest ten-thousand**   *Answer:* _____

8.   **15,448 to the nearest hundred**   *Answer:* _____

9.   **16,090 to the nearest thousand**   *Answer:* _____

10. **49,924 to the nearest hundred**   *Answer:* _____

Rounding Off Decimal Numbers

In this lesson, we will explore how to round off decimal numbers to a specified place value. You have read and heard about many different settings in which this is done, such as (a) a race car's speed, rounded off to the nearest hundredth of a mile per hour; (b) the weight of a box of cereal, rounded off to the nearest tenth of an ounce; (c) the height of a person, rounded off to the nearest tenth of a meter; and (d) the width of a pencil point, rounded off to the nearest thousandth of an inch.

Your Goal: When you have completed this lesson, you should be able to approximate any number to a specified decimal place. Remember that the purpose of rounding off decimals is to provide greater clarity while sacrificing only a small amount of accuracy in the data.

Rounding Off Decimal Numbers

LESSON 3

1

Example: *What is 0.293 rounded off to the nearest tenth?*

Solution: Beginning with the hundredths place, this digit and each placeholder to its right will be dropped. (This is similar to the method we used for counting numbers, except the digits were changed to zeros.) Now look at the hundredths place digit, which is 9. Since this digit is at least 5, the tenths digit must be increased by one. Then 0.3 is the final answer.

2

Example: *What is 0.4038 rounded off to the nearest hundredth?*

Solution: We will drop all digits to the right of the hundredths place, which are the two rightmost digits, 3 and 8. Next, look at the thousandths digit, which is 3. Since this digit is less than 5, we do <u>not</u> increase the hundredths digit. Then 0.40 is the final answer.

MathFlash!

You may wonder why we don't also drop the zero in the hundredths place to make 0.4 the final answer in Example 2. The reason is that when we round off a number to a specific decimal place, that decimal place must contain a digit, even if it is zero.

This step is unique, because for most mathematical situations, you would be encouraged to write a final answer of 0.4 instead of 0.40.

3

Example: *What is 0.16839 rounded off to the nearest thousandth?*

Solution: We must drop both the 3 and 9, since they lie to the right of the thousandths place. Next, look at the digit in the ten-thousandths place, which is 3. Since this digit is less than 5, we do not change the digit in the thousandths place, which is 8. Then 0.168 is the final answer.

4

Example: *What is 0.9516 rounded off to the nearest tenth?*

Solution: As in Example 1, the digits 5, 1, and 6 must all be dropped. We note that the digit in the hundredths place is at least 5, so this will force the digit in the tenths place to increase by one. Since the digit in the tenths place is already a 9, it will increase to a 0 and then force the 0 in the units place to become 1. Then 1.0 is the final answer. (Note that the 0 must be included in the answer. Do <u>not</u> write your answer simply as 1.)

5

Example: *What is 0.80652 rounded off to the nearest hundredth?*

Solution: The digits 6, 5, and 2 must all be dropped. Since the digit in the thousandths place (6) is greater than 5, we must increase the digit in the hundredths place (0) by one. Then 0.81 is the final answer.

6

Example: *What is 0.29985 rounded off to the nearest hundredth?*

Solution: The rightmost digits 9, 8, and 5 must all be dropped. The digit in the thousandths place is 9, so we must increase the digit in the hundredths place by one. But this digit is already a 9, so as we change this 9 to 0, we must also increase the 2 in the tenths place to 3. Then 0.30 is the final answer. Again, note that the zero in the hundredths place is required.

7

Example: *What is 0.1345 rounded off to the nearest thousandth?*

Solution: The rightmost digit, which is 5, must be dropped. Since this digit is at least 5, the digit in the thousandths place must be increased by one. Then 0.135 is the final answer.

MathFlash!

If you were asked to round off 0.1345 to the nearest hundredth, could you arrive at the correct answer? Cover up the next sentence if you feel confident! Hopefully, you realized that 0.13 is the answer.

8

Example: *What is 5.719 rounded off to the nearest tenth?*

Solution: Here we have a mixed decimal, but don't let that interfere with your thoughts. Just as before, we drop the 1 and 9. Since the digit in the hundredths place is less than 5, we don't increase the digit 7 in the tenths place. 5.7 is the final answer.

9

Example: *What is 12.9673 rounded off to the nearest hundredth?*

Solution: By now, you should be aware of the fact that the digits 7 and 3 will both drop. The digit in the thousandths place (7) is greater than 5. So, we must increase the digit in the hundredths place by one. The final answer is 12.97.

10

Example: *What is 23.00084 rounded off to the nearest thousandth?*

Solution: We will need to drop both 8 and 4. Now we inspect the digit in the ten-thousandths place, which is 8. Since this is greater than 5, we must increase the digit in the thousandths place by one. Then 23.001 is the final answer.

MathFlash!

If you were to round off the number in Example 10 to the nearest hundredth, be sure you realize that 23.00 would have been the final answer. Both zeros would need to be included in the answer.

11

Example: *What is 0.351954 rounded off to the nearest ten-thousandth?*

Solution: We need four digits to the right of the decimal point. The two rightmost digits will be dropped (5 and 4). Since the digit in the hundred-thousandths place is 5 or greater (it is in fact 5), the digit 9 must be increased by one. But this forces the digit in the thousandths place (1) to also increase by one. Then 0.3520 is the final answer.

12

Example: *What is 0.8065417 rounded off to the nearest ten-thousandth?*

Solution: We drop the digits 4, 1, and 7. Since the digit in the hundred-thousandths place (4) is less than 5, the first four digits to the right of the decimal point remain unchanged. Then 0.8065 is the final answer.

Test Yourself!

Round off each number to the specified place.

1.   3.124 to the nearest hundredth   *Answer:* _____

2.   0.8531 to the nearest tenth   *Answer:* _____

3.   0.71038 to the nearest ten-thousandth *Answer:* _____

4.   0.8961 to the nearest hundredth   *Answer:* _____

5.   0.4285 to the nearest thousandth   *Answer:* _____

6.   0.6183 to the nearest hundredth   *Answer:* _____

Test Yourself! (continued)

7.   0.0945 to the nearest tenth   *Answer:* _____

8.   0.545454 to the nearest ten-thousandth   *Answer:* _____

9.   2.17478 to the nearest thousandth   *Answer:* _____

10.   0.52983 to the nearest tenth   *Answer:* _____

11.   7.50832 to the nearest thousandth   *Answer:* _____

12.   0.11996 to the nearest ten-thousandth   *Answer:* _____

Adding/Subtracting with Signed Numbers

In this lesson, we will explore the rules that govern adding and subtracting signed numbers. There are countless examples for which signed numbers are used, such as (a) a company's year-end profit, (b) the number of yards lost on a play in football, (c) a drop in a temperature reading over a one-hour period, and (d) a person's weight gain over a week.

Your Goal: When you have completed this lesson, you should have a thorough understanding of the meaning of signed numbers.

LESSON 4

Adding/Subtracting with Signed Numbers

RULE 1   When adding two or more numbers with the same sign, simply add the numbers and attach the common sign.

1

Example: *What is the result of adding +4 to +9?*

Solution: The example can be written as (+4) + (+9). The answer is +13. We can also simply write 13. Any number without a sign is considered positive.

2

Example: *What is the result of adding −7 and −5?*

Solution: This example can be written as (−7) + (−5). The answer is −12.

3

Example: *What is the result of adding −8, −1, and −11?*

Solution: We can write this example as (−8) + (−1) + (−11). The answer is −20.

RULE 2   When adding two or more numbers with different signs, add each group of numbers with like signs, and then take the difference between the results. Mentally remove the signs. The final sign will be the sign of the number with the highest value.

4

Example: *What is the result of adding +5 and −16?*

Solution: This example can be written as (+5) + (−16). Since the difference between 5 and 16 is 11, we know that the answer is either +11 or −11. Since 16 is larger than 5, the answer is −11.

5

Example: *What is the result of adding −17, +6, +23, and −3?*

Solution: This example can be written as (−17) + (+6) + (+23) + (−3). By adding −17 to −3, we get −20. By adding +6 to +23, we get +29. Adding −20 to +29 yields the final answer of +9, or simply 9.

6

Example: *What is the result of adding –25 to the sum of 6 and –8?*

Solution: First, we will calculate 6 + (–8). We take the difference of 6 and 8 and attach the sign of the larger number; this leads to –2. Now, (–25) + (–2) = –27.

MathFlash!

The plus sign (+) serves as both the sign of a positive number and as an addition sign. For convenience and style, there are ways to reduce the number of appearances of the + sign. Here are some examples.

> *(+3) + (+7) can be written as 3 + 7.*
>
> *(–4) + (+5) can be written as –4 + 5.*
>
> *(–10) + (–2) can be written as –10 – 2.*
>
> *(–9) + (+6) + (–14) can be written as –9 + 6 – 14.*

From here on, we will use these "abbreviated" forms of writing addition (and subtraction) expressions.

RULE 3 When subtracting two numbers, change the sign of the number to be subtracted. Plus becomes minus or minus becomes plus. Then follow the rules of addition.

7

Example: *What is the result of subtracting 20 from 5?*

Solution: This is written as (+5) – (+20). Following the subtraction rule, we rewrite this as (5) + (–20), which can also be shortened to 5 – 20. Now, following the rules of addition, the answer is –15.

8

Example: *What is the result of subtracting –12 from 8?*

Solution: We can write this as 8 – (–12). Now change this to appear as 8 + (+12) = 8 + 12 = 20. (If you are losing by $12, but want to be ahead by $8, then you have to win $20.)

9

Example: *What is the result of subtracting –2 from –6?*

Solution: First, rewrite this as –6 – (–2) = –6 + (+2) = –4. –6 + 2 = –4. Notice that "+ 2" can be written instead of "+ (+2)."

10

Example: *What is the result of subtracting 18 from –7?*

Solution: We write this as (–7) – (18) = –7 –18 = –25. Note that we could have included the step before the answer as (–7) + (–18).

11

Example: *Let's try a problem that combines addition and subtraction. What is the result of subtracting –15 from the sum of 5 and –23?*

Solution: First, compute 5 + (–23) = –18. Now, –18 – (–15) = –18 + 15 = –3.

12

Example: *The temperature at 8:00 A.M. was +35°. By evening, the temperature had dropped to –12°. What is the change in temperature?*

Solution: When there is a change between two numbers, we always subtract. In this example, we must subtract +35 from –12, since a net change always means that the first number is to be subtracted from the second number. Thus, –12 – (+35) = –12 + (–35) = –47. Our final answer is that the temperature had dropped 47 degrees!

MathFlash!

Be careful when you read a word problem. If the temperature had dropped 12°, you would have been asked to find the new temperature. The correct calculation would have been +35° – 12° = +23°.

Final Thought: You should always remember that subtraction is not interchangeable. This means you cannot flip the two sides of the expression. By subtracting 4 from –2, you get a different answer than you would if subtracting –2 from 4. On the other hand, addition is interchangeable. For example, 3 + (–10) = (–10) + 3.

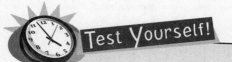

Test Yourself!

1.   What is the value of (+5) – (+9)?   *Answer:* _____

2.   What is the value of (–6) + (–4)?   *Answer:* _____

3.   What is the value of (+3) + (–9)?   *Answer:* _____

4.   What is the value of (+15) + (–7)?   *Answer:* _____

5.   What is the value of (–2) – (–9)?   *Answer:* _____

6.   What is the value of (–10) – (+11)?   *Answer:* _____

7.   What is the value of –17 subtracted
     from +12?   *Answer:* _____

8.   What is the value of –15 added to +4?   *Answer:* _____

9.   What is the value of (+32) – (+14)?   *Answer:* _____

10. What is the value of (+13) – (–3) + (–7)? *Answer:* _____

11. What is the value of (–5) + (–17) + (–4)
     + (–10)?   *Answer:* _____

12. The temperature at 9:00 A.M. was –8°.
     By late afternoon, the temperature had
     risen 31°. What was the temperature by
     late afternoon?   *Answer:* _____

Multiplying/Dividing with Signed Numbers

In this lesson, we will explore the rules that govern multiplying and dividing signed numbers. Suppose you lose $50 in a stock every week. To determine how much money you would lose in six weeks, we could consider the loss as a negative number. Then in six weeks, the amount of money that represents how much less money you will have is shown by the multiplication (–50)(6).

Also, if the temperature has been dropping at a rate of 2 degrees per hour for the past five hours, then we can use the number –2 as the change in temperature for each hour. Thus, if we wanted to know how much higher the temperature was three hours ago, compared to the temperature now, we would write (–2)(–3).

Your Goal: When you have completed this lesson, you should have a complete understanding of all four basic operations with signed numbers (addition, subtraction, multiplication, and division).

LESSON 5

Multiplying/Dividing with Signed Numbers

When multiplying numbers, the result is called the **product**.

RULE 1 When multiplying two numbers, $(+)(+) = +$, $(+)(-) = -$, and $(-)(-) = +$.

1

Example: *What is the value of (5)(–6)?*

Solution: The answer is –30. Note the absence of the + sign in front of the 5.

2

Example: *What is the value of (–12)(–4)?*

Solution: The answer is 48.

3

Example: *What is the value of (17)(4)?*

Solution: The answer is 68. Note the absence of the + sign in front of all these numbers. As before, they are assumed positive when no sign precedes them.

RULE 2 When multiplying three or more numbers, if they are all positive, the product is positive. If there is a mixture of signs or if they are all negative, simply count the number of negative signs. If this number of negative signs is odd (1, 3, 5, etc.), then the product is negative. Otherwise, the product is positive.

4

Example: *What is the value of (4)(–3)(–2)(–6)?*

Solution: The answer is –144.

5

Example: *What is the value of (2)(–10)(–5)(–6)(–3)?*

Solution: The answer is 1800.

MathFlash!

When any product contains the number zero, the result must be zero. So, (–6)(7)(0) = 0 and (–1)(–2)(–3)(0) = 0.

RULE 3   When dividing two numbers, follow the rules of multiplying the two numbers. The answer to a division problem is called the **quotient**.

6

Example: *What is the value of 14 ÷ (–2)?*

Solution: The answer is –7.

7

Example: *What is the value of (–50) ÷ (–5)?*

Solution: The answer is 10.

8

Example: *What is the value of (–60) ÷ (+10) ÷ (–2)?*

Solution: (–60) ÷ (+10) = –6. Then (–6) ÷ (–2) = 3. It is extremely important that you do the actual divisions in the order that they appear. If you divide +10 by –2 first, you will get –5. Then –60 divided by –5 equals 12, which is a wrong answer.

MathFlash!

Zero divided by a nonzero number is zero. However, any number (including zero) that is divided by zero is undefined (has no answer).

9

Example: *What is the value of 0 ÷ –54?*

Solution: 0 ÷ (–54) = 0. Zero divided by any nonzero number has an answer of zero.

10

Example: *What is the value of 0 ÷ 0?*

Solution: The answer is undefined, since we are dividing by zero.

11

Example: *What is the value of 32 ÷ 0?*

Solution: The answer is undefined, since we are dividing by zero.

 **Test Yourself!**

1. What is the value of (+7)(–4)? *Answer:* _____

2. What is the value of (–5)(–11)? *Answer:* _____

3. What is the value of (–3)(+14)(–2)(–5)? *Answer:* _____

4. What is the value of (+90) ÷ (–15)? *Answer:* _____

5. What is the value of (+40) ÷ (+8)? *Answer:* _____

6. What is the value of (–28) ÷ (–4)? *Answer:* _____

7. What is the value of 96 ÷ (–8) ÷ (2)? *Answer:* _____

8. Which one of the following is undefined?
    (A) 20 ÷ (–20)   (C) (–20)(0)
    (B) 20 ÷ 0   (D) 0 ÷ 20   *Answer:* _____

9. What is the value of (2000)(–300)(0)? *Answer:* _____

10. If a person earns $800 per week,
    how much money will that
    person have earned in 12 weeks? *Answer:* _____

LESSONS 1-5

QUIZ ONE

1. Which one of the following is the correct wording for the number 0.0915?

   A Nine hundred fifteen hundredths

   B Nine and fifteen thousandths

   C Nine hundred fifteen ten-thousandths

   D Nine hundred fifteen thousandths

2. How is the expression "seven hundred two thousand sixty" written in digits?

   A 702,060

   B 702,600

   C 720,060

   D 720,600

3. What is the value of (–25) – (–11)?

   A –36

   B –14

   C 14

   D 36

4. The elevation of Death Valley in California is 282 feet below sea level, which is represented as –282 feet. The elevation of Black Mountain in Kentucky is 750 feet above sea level, which is represented as +750 feet. How many feet higher in elevation is Black Mountain than Death Valley?

   A 234

   B 468

   C 516

   D 1032

5. Which one of the following is the correct wording for the number 40,809?

   A Forty thousand eight hundred ninety

   B Forty thousand eight hundred nine

   C Four thousand eight hundred nine

   D Four thousand eighty-nine

6. How is the expression "two hundred six thousandths" written in digits?

   A 0.260

   B 0.206

   C 0.026

   D 0.0206

7. If +17 is subtracted from the sum of –6 and +11, what is the result?

   A 22

   B 12

   C –12

   D –22

8. **What is the value of 0 ÷ 0?**

    A −1

    B 0

    C 1

    D It is undefined.

9. **What is the number 0.2897 rounded off to the nearest hundredth?**

    A 0.29

    B 0.3

    C 0.30

    D 0.28

10. **What is the number 26,475 rounded off to the nearest thousand?**

    A 26,000

    B 26,400

    C 26,500

    D 27,000

Reducing Fractions to Lowest Terms

In this lesson, we will explore how to reduce any fraction to its lowest terms. There are many instances in which the use of a reduced fraction simplifies a concept. For example, suppose you read that out of 245 doctors that are surveyed, 98 of them need additional staff. As fractions really represent ratios, it is not easy to use the comparison of these numbers. However, if you knew how to simplify $\frac{98}{245}$ to $\frac{2}{5}$, it is much easier to understand this example. We can picture a group of five doctors of which (on the average) two of them need the additional staff.

You will also learn how to reduce fractions in which units are not similar. For example, the numerator of the fraction may have seconds as its units, whereas the denominator may have hours.

Your Goal: When you have completed this lesson, you should be able to reduce any fraction to its lowest terms, even if the units expressed are different for the numerator and denominator. It will be important to remember that sometimes, due to the size of the numbers, two or three steps may be required to be sure that the fraction is completely reduced.

Reducing Fractions to Lowest Terms

Fractions such as $\frac{1}{2}$ and $\frac{11}{3}$ are considered reduced to lowest terms. For each fraction, there is no single number (except for 1) that can divide into both the numerator and denominator. But the fractions $\frac{6}{9}$ and $\frac{28}{21}$ each have at least one number (besides 1) that can be divided into both the numerator and denominator. For the fraction $\frac{6}{9}$, the number 3 is a factor of both 6 and 9; for the fraction $\frac{28}{21}$, the number 7 is a factor of both 28 and 21.

1

Example: *What is the fraction $\frac{8}{12}$ in lowest terms?*

Solution: We first determine that 4 is a common factor of 8 and 12.

Next, divide both 8 and 12 by 4. Since $12 \div 4 = 3$ and $8 \div 4 = 2$, the answer is $\frac{2}{3}$.

Had you simply divided the numerator and denominator by 2, the fraction would appear as $\frac{4}{6}$. Then, you would need to divide the new numerator and denominator by 2 to arrive at the correct answer.

2

Example: *What is the fraction $\frac{35}{10}$ in lowest terms?*

Solution: We first determine that 5 is a common factor of 10 and 35, which means that you can divide both 10 and 35 by 5 and have no remainder. Now divide 35 by 5, and also divide 10 by 5. Since $35 \div 5 = 7$ and $10 \div 5 = 2$, the answer is $\frac{7}{2}$.

Even though the answer is an improper fraction (the numerator is larger than the denominator), it is still considered reduced, provided there is no number that can divide into both the numerator and the denominator. Be aware that the mixed fraction $3\frac{1}{2}$ would also be correct. A mixed fraction is a combination of a whole number and a proper fraction (the numerator is smaller than the denominator).

Example: *What is the fraction $\frac{7}{56}$ in lowest terms?*

Solution: We first notice that both the numerator and denominator are divisible by 7. After dividing 7 by 7 and 56 by 7, we get 1 and 8, in turn. Thus, the answer is $\frac{1}{8}$.

Example: *What is the fraction $\frac{36}{9}$ in lowest terms?*

Solution: Here, 9 is a factor of both the numerator and denominator, so after dividing each of these by 9, the answer is $\frac{4}{1}$. Normally, when the denominator becomes 1, the final answer is written without the denominator. So, in this example, we could simply write the answer as 4.

Example: *What is the fraction $\frac{111}{156}$ in lowest terms?*

Solution: Since both the numerator and denominator are odd, we should try to use an odd number as a divisor, which means a common factor. Since 3 can divide into both the numerator and denominator, we do the division by 3 to get $\frac{37}{52}$.

To be sure that this is the correct answer, we observe that 37 and 52 share no common divisors.

6

Example: *What is the fraction $\dfrac{216}{112}$ in lowest terms?*

Solution: Although you may immediately recognize that both numbers are divisible by 2 (since they are even), you might also see that the number 4 divides into these two numbers. So divide the numerator and denominator by 4 to get $\dfrac{54}{28}$. We need to go another step! Each of these numbers is divisible by 2, so the final answer is $\dfrac{27}{14}$.

The only way to have completed this example in one step is to have divided the numerator and denominator by 8.

7

Example: *What is the fraction $\dfrac{65}{215}$ in lowest terms?*

Solution: Examples such as this are really friendly because both numbers end in 5. We know that 5 must be a divisor of both the numerator and denominator. Sure enough, when we divide both parts of the fraction by 5, our answer becomes $\dfrac{13}{43}$.

8

Example: *What is the fraction $\dfrac{119}{140}$ in lowest terms?*

Solution: This may look like "Mission Impossible" since 119 does not have any obvious divisors. For example, you know that 10 is a divisor of the denominator, but it is certainly not a divisor of the numerator. In cases such as this, don't despair. Be prepared to simply try numbers like 7, 11, or 13. The good news is that 7 is a divisor of both the numerator and denominator. Upon dividing the numerator and denominator by 7, we get the answer of $\dfrac{17}{20}$.

Example: *What is the ratio* $\dfrac{4\ minutes}{6\ hours}$ *in lowest terms?*

9

Solution: Notice that we have units associated with these numbers, so we cannot simply reduce $\dfrac{4}{6}$. We first need to convert either or both the numerator and denominator to a common unit. Here, we will change 6 hours to (6)(60) = 360 minutes. Then we have the same units on the top and bottom of this fraction, so they "cancel out." We only have to reduce $\dfrac{4}{360}$, which becomes $\dfrac{1}{90}$.

Example: *What is the ratio* $\dfrac{15\ inches}{3\ feet}$ *in lowest terms?*

10

Solution: If you didn't notice the difference in units, you could mistakenly write $\dfrac{5}{1}$ or 5 as your answer. Let's change the 3 feet to (3)(12) = 36 inches. Now we must reduce $\dfrac{15}{36}$. We can spot 3 as a common divisor, so the answer is $\dfrac{5}{12}$.

MathFlash!

Not every fraction requires reduction. Fractions such as $\dfrac{3}{11}$ and $\dfrac{20}{9}$ cannot be reduced any further. Remember that in order to reduce a given fraction to lowest terms, you must first find a common factor (except 1) for both the numerator and denominator. If no such factor exists, then the fraction is already in its reduced form.

Example: *What is the ratio* $\dfrac{8\ pounds}{24\ ounces}$ *in lowest terms?*

11

Solution: Once again, it would be an error to simply reduce $\dfrac{8}{24}$ to $\dfrac{1}{3}$! Since there are 16 ounces in one pound, 8 pounds becomes (8)(16) = 128 ounces. So, $\dfrac{128}{24}$ becomes the ratio to reduce. By using 8 as a common divisor, the answer becomes $\dfrac{16}{3}$.

1.   What is $\dfrac{12}{20}$ in reduced fraction form?

      Answer: _____

2.   What is $\dfrac{64}{56}$ in reduced fraction form?

      Answer: _____

3.   What is $\dfrac{30}{78}$ in reduced fraction form?

      Answer: _____

4.   What is $\dfrac{27}{207}$ in reduced fraction form?

      Answer: _____

5.   What is $\dfrac{75}{55}$ in reduced fraction form?

      Answer: _____

6.   What is $\dfrac{126}{161}$ in reduced fraction form?

      Answer: _____

7.   **What is the ratio of 5 inches to 10 feet in reduced form?**

      Answer: _____

8.   **What is the ratio of 4 pounds to 7 ounces in reduced form?**

      Answer: _____

9.   **What is the ratio of 18 minutes to 2 hours in reduced form?**

      Answer: _____

10.   **What is the ratio of 8 months to 9 years in reduced form?**

      Answer: _____

11.   **What is the ratio of 2 years to 13 weeks in reduced form?**

      Answer: _____

12.   **What is the ratio of 10 feet to 6 yards in reduced form?**

      Answer: _____

Here is an interesting observation: Think of any two consecutive numbers, such as 3 and 4. If you write a fraction using one of these numbers in the numerator and the other in the denominator, the fraction is automatically reduced to lowest terms! You should try this with different pairs of consecutive numbers.

Converting Improper Fractions to Mixed Fractions and Vice Versa

In this lesson, we will look at the various equivalent forms of a fraction. Specifically, we'll investigate improper fractions and mixed fractions. In most cases, the mixed fraction would be the desired form; for example, you might include $3\frac{1}{4}$ teaspoons of sugar in a recipe for baking a cake. There are times when an improper fraction is used; for example, a teacher might need to report the ratio of girls to boys in a class. If there are 18 girls and 11 boys, the ratio can be written as 18:11 or as the fraction $\frac{18}{11}$.

Your Goal: When you have completed this lesson, you should be able to change any improper fraction to a mixed fraction and vice versa.

Converting Improper Fractions to Mixed Fractions and Vice Versa

An **improper fraction** is one in which the numerator (top) is larger than the denominator (bottom). Examples are $\frac{17}{11}$ and $\frac{24}{5}$. A **mixed fraction** is one in which there is a whole number part and a proper fraction part. In **proper fractions**, the numerator is smaller than the denominator. Examples of mixed fractions are $4\frac{1}{2}$ and $10\frac{3}{7}$.

> To change an improper fraction to a mixed fraction, simply divide the denominator into the numerator to get a quotient of some whole number.
> The remainder of this division is then placed on top of the original denominator to form the proper fraction part of the answer.
> Finally, check to be sure that the proper fraction is in reduced form.

1

Example: *What is the mixed fraction form of $\frac{10}{3}$?*

Solution: Upon dividing 10 by 3, the quotient is 3, and the remainder is 1. The whole number from the quotient becomes the whole number part of the answer, and the remainder is placed over the original denominator to become the proper fraction part of the answer.

This means that the answer is $3\frac{1}{3}$.

2

Example: *What is the mixed fraction form of $\frac{39}{5}$?*

Solution: Upon dividing 39 by 5, the quotient is 7, and the remainder is 4. This means that the answer is $7\frac{4}{5}$. Again, notice how the remainder is included in the answer.

3

Example: *What is the mixed fraction form of $\frac{34}{4}$?*

Solution: Upon dividing 34 by 4, the quotient is 8, and the remainder is 2. This means that the answer would appear to be $8\frac{2}{4}$. But, as was often said in the old Western movies, "Hold your horses!" We cannot leave an answer as $\frac{2}{4}$. Since 2 divides into both the numerator and denominator, we must reduce this fraction to $\frac{1}{2}$. The final answer is $8\frac{1}{2}$.

4

Example: *What is the mixed fraction form of $\frac{12}{9}$?*

Solution: Upon dividing 12 by 9, the quotient is 1, and the remainder is 3. Initially, the answer would be $1\frac{3}{9}$, but you should notice that 3 divides into both the numerator and denominator. For this reason, the final answer is $1\frac{1}{3}$.

5

Example: *What is the mixed fraction form of $\frac{28}{7}$?*

Solution: When you divide 28 by 7, the quotient is 4, and there is no remainder. This means that there is no proper fraction left, so the final answer is 4. You would not write $4\frac{0}{4}$. Technically, we have changed a mixed fraction into a whole number, not into a mixed fraction. Do not be too concerned that there is no remainder. In most cases where this type of example appears, the instructions will be to simplify the fraction.

Now, let's show how to reverse this process. **To change a mixed fraction into an improper fraction,** we multiply the denominator of the proper fraction by the whole number, and then add the numerator. This becomes the new numerator, and the original denominator is left alone. As before, the last item to check is that the improper fraction is in reduced form.

6

Example: *What is the improper fraction form of $5\frac{1}{8}$?*

Solution: Multiply 8 by 5, and then add 1. This will yield 41. Then, by placing 41 over the original denominator, the final answer is $\frac{41}{8}$.

7

Example: *What is the improper fraction form of $3\frac{3}{10}$?*

Solution: Multiply 10 by the whole number 3, and then add the 3 in the numerator. This will yield 33. Then, 33 is placed over 10, so that the final answer is $\frac{33}{10}$.

8

Example: *What is the improper fraction form of $6\frac{2}{14}$?*

Solution: Following our basic rule, we multiply 14 by 6 and then add 2. This will yield 86. By placing 86 over 14, it would appear that our answer is $\frac{86}{14}$. Like an unfinished painting, this needs some finishing touches. Hopefully, you can see that 2 divides into both the numerator and denominator, so that the final answer is $\frac{43}{7}$.

Can you spot an easier way to handle this example? Look back at the original mixed fraction. The proper fraction part, namely, $\frac{2}{14}$, could have been reduced to $\frac{1}{7}$ first. Then, we would have been asked to change $6\frac{1}{7}$ into an improper fraction. Following the rules, multiply 7 by 6, and then add 1. This will yield 43, and then by placing 43 over 7, the final answer is $\frac{43}{7}$.

MathFlash!

*Whenever you see an example such as Example 8, it is usually easier to **first reduce the proper fraction to lowest terms**.*

9

Example: *What is the improper fraction form of $2\frac{16}{48}$?*

Solution: We will save time by first noticing that 16 divides into the numerator and denominator. Thus, we can write the original mixed fraction as $2\frac{1}{3}$. Multiply 3 by 2 and add 1 to get 7. Then 7 is placed over 3 to give the final answer of $\frac{7}{3}$.

10

Example: *What is the improper fraction form of $9\frac{24}{33}$?*

Solution: First, let's reduce the mixed fraction to $9\frac{8}{11}$, since 3 is a factor of both 24 and 33. Now, multiply 11 by 9, and add 8 to get 107. The final answer is $\frac{107}{11}$.

11

Example: *In a large room, there are 76 women and 52 men. What is the ratio of women to men in reduced improper fraction form?*

Solution: The initial ratio appears as $\frac{76}{52}$. Then, since both the numerator and denominator can be divided by 4, the reduced fraction is $\frac{19}{13}$, which is the final answer.

12

Example: *Return to Example 11. What is the mixed fraction form (reduced) of the ratio of women to men?*

Solution: This is equivalent to changing $\frac{19}{13}$ into a mixed fraction. The final answer is $1\frac{6}{13}$.

Test Yourself!

1. What is the mixed fraction form of $\frac{17}{9}$?   *Answer:* _____

2. What is the mixed fraction form of $\frac{26}{6}$?   *Answer:* _____

3. What is the mixed fraction form of $\frac{99}{12}$?   *Answer:* _____

4. What is the mixed fraction form of $\frac{134}{20}$?   *Answer:* _____

5. What is the mixed fraction form of $\frac{89}{17}$?   *Answer:* _____

6. What is the improper fraction form of $3\frac{2}{7}$?   *Answer:* _____

7. What is the improper fraction form of $7\frac{4}{11}$?   *Answer:* _____

8. What is the improper fraction form of $4\frac{9}{10}$?   *Answer:* _____

9. What is the improper fraction form of $5\frac{4}{18}$?   *Answer:* _____

10. What is the improper fraction form of $8\frac{28}{35}$?   *Answer:* _____

11. In a certain zoo, there are 128 lions and 40 tigers. What is the ratio of lions to tigers in reduced improper fraction form?

    Answer: _____

12. Return to #11. What is the ratio of lions to tigers in mixed fraction form?

    Answer: _____

Adding/Subtracting Fractions

In this lesson, we will explore how to add and subtract fractions. As you will find out, it is important to pay attention to the denominators of the given fractions. When denominators are different, there are a number of additional steps to add or subtract fractions than if the denominators are the same. For example, suppose you bought a stock last week and found out this week that the value of this stock rose by $\frac{1}{8}$ point. This week the stock is doing so well that it rises by $\frac{3}{4}$ point. In order to see how many points this stock has risen since you purchased it, you would have to add $\frac{1}{8}$ and $\frac{3}{4}$.

Your Goal: When you have completed this lesson, you should be able to add and subtract fractions, whether or not their denominators are the same. Remember to check that your answer is in reduced form.

LESSON 8

Adding/Subtracting Fractions

RULE 1   When adding or subtracting fractions with the same denominator, simply add or subtract the numerators. The resulting fraction will have the same denominator as the original fractions.

1

Example: *What is the value of $\frac{2}{5} + \frac{7}{5}$?*

Solution: The denominators are equal, so simply add 2 to 7 to get the correct numerator. The answer is $\frac{9}{5}$. As discussed in a previous lesson, this fraction is considered reduced since there is no number that can divide into both 9 and 5. If you wish, you may also write the answer as a mixed number, $1\frac{4}{5}$.

2

Example: *What is the value of $\frac{4}{3} - \frac{11}{3} + \frac{1}{3}$?*

Solution: Again, the denominators are equal, so we calculate $4 - 11 + 1 = -6$. The answer starts as $\frac{-6}{3}$, but must be reduced to -2.

RULE 2   When adding or subtracting fractions with different denominators, each fraction must be changed to an equivalent fraction using the common denominator for all given fractions. Then proceed as in Rule 1.

3

Example: *What is the value of $\frac{1}{2} + \frac{5}{6}$?*

Solution: Using our knowledge of common denominators, 6 is the number to use. Convert $\frac{1}{2}$ to $\frac{3}{6}$ so that the problem can be written as $\frac{3}{6} + \frac{5}{6} = \frac{8}{6}$, which should be reduced to $\frac{4}{3}$.

Example: *What is the value of $-\frac{3}{4} + \frac{7}{12} - \frac{9}{8}$?*

4

Solution: First, recognize that any number that is divisible by 8 is also divisible by 4. Thus, we need a common denominator for 12 and 8. That number is 24. Change all three fractions to ones with 24 as the denominator. Rewrite as $-\frac{18}{24} + \frac{14}{24} - \frac{27}{24} = -\frac{31}{24}$. Note that this answer is considered reduced to lowest terms since 31 and 24 share no common divisor (except 1).

MathFlash!

If you decided to use a larger number than 24, you could still get the correct answer. Let's suppose you decided to use 48 since all three denominators divide into 48. Then you would have $-\frac{36}{48} + \frac{28}{48} - \frac{54}{48} = -\frac{62}{48}$, which reduces to $-\frac{31}{24}$.

Example: *What is the value of $\frac{19}{7} - \frac{3}{2} - \frac{1}{3}$?*

5

Solution: The easiest way to find a common denominator in this problem is to simply multiply the three numbers: $(7)(2)(3) = 42$, which is the least common denominator. After changing each fraction to one with a denominator of 42, the problem appears as $\frac{114}{42} - \frac{63}{42} - \frac{14}{42} = \frac{37}{42}$.

Example: *What is the value of $2\frac{3}{4} + 5\frac{1}{2}$?*

6

Solution: Change the second fractional part so that 4 becomes the denominator. The problem then reads as $2\frac{3}{4} + 5\frac{2}{4} = 7\frac{5}{4}$. Now recall that a mixed number cannot be left in which the numerator exceeds the denominator, so change $\frac{5}{4}$ to $1\frac{1}{4}$. Once this is added to 7, the final answer becomes $8\frac{1}{4}$.

7

Example: *What is the value of $9\frac{3}{8} - 3\frac{4}{5}$?*

Solution: Change both fractional parts to a common denominator of 40.
Since $\frac{3}{8} = \frac{15}{40}$ and $\frac{4}{5} = \frac{32}{40}$, the problem reads as $9\frac{15}{40} - 3\frac{32}{40}$.
Now, we need to change the first of these two mixed fractions so that it contains more fortieths than the second fraction. Since $1 = \frac{40}{40}$, we can express $9\frac{15}{40}$ as $8 + 1 + \frac{15}{40} = 8 + \frac{40}{40} + \frac{15}{40} = 8\frac{55}{40}$.
Then, $8\frac{55}{40} - 3\frac{32}{40} = 5\frac{23}{40}$.

8

Example: *What is the value of $10\frac{8}{9} - 2\frac{1}{6}$?*

Solution: Change both fractional parts to a common denominator of 18.
$\frac{8}{9} = \frac{16}{18}$ and $\frac{1}{6} = \frac{3}{18}$. Thus, the problem can be rewritten as $10\frac{16}{18} - 2\frac{3}{18} = 8\frac{13}{18}$.

MathFlash!

In any of these last three examples, another approach would be to work with improper fractions. Thus, in Example 7, we could have written $\frac{75}{8} - \frac{19}{5}$. Changing each fraction so that the denominator is 40, we get $\frac{375}{40} - \frac{152}{40} = \frac{223}{40}$. If you use this approach, the fraction $\frac{223}{40}$ is acceptable as the final answer. This is equivalent to $5\frac{23}{40}$.

Example: *What is the value of $3\frac{3}{4} - 5\frac{5}{6} + 7\frac{7}{9}$?*

9

Solution: Let's change each of these to improper fractions. Then, the example would read as $\frac{15}{4} - \frac{35}{6} + \frac{70}{9}$. A common denominator for 4, 6, and 9 can be found by simply taking multiples of 9 until reaching a number that is divisible by both 4 and 6. This number is 36. Change each fraction to one that has 36 as its denominator. We then have $\frac{135}{36} - \frac{210}{36} + \frac{280}{36} = \frac{205}{36}$. Since 36 and 205 do not share a common divisor (except the number 1), this answer is considered reduced to lowest terms.

If you prefer the mixed fraction form, the answer would be $5\frac{25}{36}$.

MathFlash!

When these rules are applied to mixed numbers, it is usually easiest to leave the whole number part alone until the end of the problem, unless there is "borrowing" when doing subtraction. Of course, you still need common denominators.

Example: *What is the value of $11\frac{1}{12} + 7\frac{1}{8} - 3\frac{7}{16}$?*

10

Solution: We plan to initially use improper fractions. At this stage, the easiest way to get a common denominator is to take multiples of 16 until you find a number that is divisible by 12 and by 8. With just a little patience, you should be able to determine that 48 is a good candidate. Rewrite the problem as $11\frac{4}{48} + 7\frac{6}{48} - 3\frac{21}{48}$. It will be quicker to now change each fraction into improper form, so that the problem reads $\frac{532}{48} + \frac{342}{48} - \frac{165}{48} = \frac{709}{48}$.

Alternatively, you could add the first two fractions to get $18\frac{10}{48}$.

Now we are faced with subtracting $3\frac{21}{48}$ from $18\frac{10}{48}$. Since 10 is less

than 21, we must change $18\frac{10}{48}$ to $17\frac{58}{48}$, so that the subtraction

reads $17\frac{58}{48} - 3\frac{21}{48} = 14\frac{37}{48}$. Fortunately, this is probably the most

difficult addition/subtraction problem with fractions you will ever

meet.

Example: *Jaclyn has $1\frac{5}{6}$ gallons of paint. If she uses $\frac{3}{5}$ gallon to paint a room,*
11 *how much paint is left over?*

Solution: We must subtract $\frac{3}{5}$ from $1\frac{5}{6}$. Then, $1\frac{5}{6} - \frac{3}{5} = \frac{11}{6} - \frac{3}{5}$. Changing
each fraction to a common denominator of 30, we have:
$\frac{55}{30} - \frac{18}{30} = \frac{37}{30}$ or $1\frac{7}{30}$ gallons of paint left over.

Example: *Peter has discovered that he spends $\frac{3}{10}$ of his weekly paycheck*
12 *on rent and $\frac{1}{12}$ of his weekly paycheck on entertainment. What*
fraction of his weekly paycheck does he spend on rent and
entertainment combined?

Solution: In this example, we must add these fractions. Then,
$\frac{3}{10} + \frac{1}{12} = \frac{18}{60} + \frac{5}{60} = \frac{23}{60}$.

1. What is the value of $\frac{1}{8} + \frac{3}{8}$?   *Answer:* _____

2. What is the value of $\frac{1}{3} + \frac{3}{5}$?   *Answer:* _____

3. What is the value of $\frac{9}{7} - \frac{3}{4}$?   *Answer:* _____

4. What is the value of $\frac{3}{2} + \frac{7}{8} - \frac{5}{6}$?   *Answer:* _____

5. What is the value of $\frac{3}{4} + \frac{5}{14}$?   *Answer:* _____

6. What is the value of $\frac{13}{15} - \frac{5}{6}$?   *Answer:* _____

7. What is the value of $\frac{7}{3} + \frac{8}{9} - \frac{1}{12}$?   *Answer:* _____

8. What is the value of $3\frac{5}{7} + 2\frac{4}{5}$?   *Answer:* _____

9. What is the value of $5\frac{2}{3} - 2\frac{7}{8}$?   *Answer:* _____

10. What is the value of $6\frac{3}{4} + 2\frac{4}{5} - 1\frac{7}{10}$?   *Answer:* _____

11. Katherine has completed $\frac{5}{16}$ of her science project. Her goal is to have $\frac{3}{4}$ of this project finished by today. What fraction of this project must she still complete in order to reach her goal?

Answer: _____

12. Ken has mowed $\frac{5}{9}$ of his lawn. His friend Linda has offered to mow $\frac{2}{5}$ of his lawn. What fraction of the lawn will have been mowed altogether?

Answer: _____

Multiplying/Dividing Fractions

In this lesson, we will explore how to multiply and divide fractions. There may be several ways to arrive at the correct reduced fraction; namely (a) reduce the size of the given numerators/denominators before performing the multiplication (or division) or (b) perform the multiplication (or division) first, then reduce the fraction, if necessary. For most examples, either method will work equally well. For instance, if you have a bag of oranges that weighs $3\frac{2}{5}$ pounds, but you only need $\frac{3}{4}$ of a bag, the weight, in pounds, of $\frac{3}{4}$ of a bag would be found by multiplying $\frac{3}{4}$ by $3\frac{2}{5}$.

Your Goal: When you have completed this lesson, you should be able to multiply and divide fractions. Remember to always check that your answer is in reduced form.

LESSON 9

Multiplying/Dividing Fractions

RULE 1 When multiplying fractions, multiply all numerators and multiply all denominators, then reduce to lowest terms.

MathFlash!

Before multiplying numerators or denominators, you may "replace" any numerator and any denominator with a smaller number, if they have a common factor. This technique will be shown in some of the examples that follow.

1

Example: *What is the value of $\frac{2}{3} \times \frac{4}{7}$?*

Solution: In checking the numerators and denominators, we note that there is no common factor. Simply multiply the numerators and then the denominators to get $\frac{8}{21}$.

2

Example: *What is the value of $\frac{8}{5} \times \frac{1}{10}$?*

Solution: Here, we notice that 8 and 10 have a common factor of 2. By dividing each of these numbers by 2, the problem appears as $\frac{4}{5} \times \frac{1}{5}$, so the answer is $\frac{4}{25}$. Note that you are <u>not allowed</u> to divide <u>only</u> denominators or <u>only</u> numerators!

Example: *What is the value of $\dfrac{6}{35} \times \dfrac{49}{6}$?*

3

Solution: First notice that 6 appears in both the numerator and the denominator. Whenever this happens, these 6's can be effectively "cancelled." Now we have $\dfrac{1}{35} \times \dfrac{49}{1}$. Also, 35 and 49 have a common factor of 7, so divide each of these numbers by 7. Now, the problem becomes $\dfrac{1}{5} \times \dfrac{7}{1} = \dfrac{7}{5}$.

MathFlash!

You can have multiplication with more than two fractions. However, you won't see it very often.

Example: *What is the value of $\dfrac{5}{8} \times \dfrac{7}{15} \times \dfrac{2}{21}$?*

4

Solution: You can choose to just multiply all numerators and all denominators and then reduce to lowest terms at the end. The fraction would appear as $\dfrac{70}{2520}$. You could now strike out the end zeros, because this is the same as dividing by 10. The fraction now reads as $\dfrac{7}{252}$. Finally, you can check that 7 is a factor of 252, so the final answer is $\dfrac{1}{36}$. Caution: Do not throw away the 1 from the numerator of the answer!

Or you could solve this problem by reducing the size of the original numbers. Dividing by 5, the 5 becomes 1, and the 15 becomes 3. Dividing the 2 and 8 by 2, the numbers become 1 and 4. Also, dividing by 7, the numbers 7 and 21 become 1 and 3.

Then $\dfrac{\overset{1}{\cancel{5}}}{\underset{4}{\cancel{8}}} \times \dfrac{\overset{1}{\cancel{7}}}{\underset{3}{\cancel{15}}} \times \dfrac{\overset{1}{\cancel{2}}}{\underset{3}{\cancel{21}}} = \dfrac{1}{4} \times \dfrac{1}{3} \times \dfrac{1}{3} = \dfrac{1}{36}$.

Example: *Let's try another multiplication problem with three fractions, and to make life more interesting, we'll use a mixed fraction in this group. What is the value of $\frac{3}{8} \times 1\frac{7}{9} \times \frac{1}{5}$?*

5

Solution: Changing the mixed fraction to an improper fraction, we have $\frac{3}{8} \times \frac{16}{9} \times \frac{1}{5}$. Then we can divide 3 into 3 to get 1, and 3 into 9 to get 3, so that the problem would reduce to $\frac{1}{8} \times \frac{16}{3} \times \frac{1}{5}$.

Next, divide 8 into 8 to get 1, and divide 8 into 16 to get 2, so that we have $\frac{1}{1} \times \frac{2}{3} \times \frac{1}{5}$.

Now, simply multiply as usual to get the answer of $\frac{2}{15}$.

RULE 2   When dividing fractions, invert the fraction(s) that follow the division sign, and then follow Rule 1.

MathFlash!

*The process of inverting the numerator and denominator in a fraction essentially means "flipping" the fraction. Thus, a fraction such as $\frac{2}{3}$ would become $\frac{3}{2}$. The number $\frac{3}{2}$ is called the **reciprocal** of $\frac{2}{3}$.*

Example: *What is the value of $\frac{3}{10} \div \frac{7}{9}$?*

6

Solution: First, flip the fraction that follows the division sign. Then rewrite this as a multiplication problem: $\frac{3}{10} \times \frac{9}{7} = \frac{27}{70}$.

Example: *What is the value of $\dfrac{5}{2} \div \dfrac{20}{21}$?*

7

Solution: As in Example 6, first rewrite this as $\dfrac{5}{2} \times \dfrac{21}{20}$. We could simply multiply numerators and denominators. But 5 can be divided into both the 5 and 20. The multiplication problem will then appear as $\dfrac{1}{2} \times \dfrac{21}{4}$, which equals $\dfrac{21}{8}$. Remember that answers may be left in improper form, if they are reduced to lowest terms.

Example: *Let's show an example involving two divisions. What is the value of $\dfrac{8}{5} \div \dfrac{1}{9} \div \dfrac{12}{7}$?*

8

Solution: In this problem, both the second and third fractions must be inverted. Now, the problem appears as $\dfrac{8}{5} \times \dfrac{9}{1} \times \dfrac{7}{12}$. Notice that both 8 and 9 have common factors with 12. Focusing on the 8 and 12, they share a common factor of 4. Dividing each of these numbers by 4, the problem would simplify to $\dfrac{2}{5} \times \dfrac{9}{1} \times \dfrac{7}{3}$.

Now, see that 3 can divide into both 3 and 9. This simplifies the problem to $\dfrac{2}{5} \times \dfrac{3}{1} \times \dfrac{7}{1}$, which equals $\dfrac{42}{5}$.

Example: *Of course mixed fractions can occur in division problems, just as they did in multiplication problems. What is the value of $2\dfrac{7}{8} \div 4\dfrac{2}{7}$?*

9

Solution: First change each fraction to its improper form. The problem will appear as $\dfrac{23}{8} \div \dfrac{30}{7}$. Next rewrite this as a multiplication, $\dfrac{23}{8} \times \dfrac{7}{30}$. Since no cancellation is possible, just multiply the numerators and denominators to get the answer of $\dfrac{161}{240}$.

Be careful! You cannot cancel or reduce numbers that are entirely in the numerator or entirely in the denominator. One number has to be a numerator and one a denominator. So, in an example such as $\frac{5}{9} \times \frac{10}{11}$, you cannot divide 5 into 5 and 5 into 10. The answer for such an example is simply $\frac{50}{99}$.

Example: *What is the value of $4\frac{1}{3} \div \frac{4}{11} \times \frac{15}{26}$?*

10

Solution: Change the first fraction into improper form and, at the same time, invert the second fraction. The problem would then appear as $\frac{13}{3} \times \frac{11}{4} \times \frac{15}{26}$. Now, see that 13 can divide into both 13 and 26. Also, 3 can divide into both 3 and 15. When all this is done, the problem will simplify to $\frac{1}{1} \times \frac{11}{4} \times \frac{5}{2}$. If you wish to, you may simply drop the first of these three fractions. Now multiply $\frac{11}{4} \times \frac{5}{2}$ to get the answer of $\frac{55}{8}$.

Example: *Lenny has written $\frac{1}{4}$ of his term paper. Also, $\frac{2}{5}$ of what he has already written consists of real-life examples. If he does not use any more real-life examples in his term paper, what fraction of his paper consists of real-life examples?*

11

Solution: In this example, the expression "$\frac{2}{5}$ of" means "$\frac{2}{5}$ times," so we simply multiply the two fractions. Then, $\frac{1}{4} \times \frac{2}{5} = \frac{2}{20} = \frac{1}{10}$.

Example: *Janice has $9\frac{1}{3}$ pounds of cashew nuts. She wants to put an equal number of pounds of these nuts into six cans. How many pounds of cashew nuts will be put into each can?*

12

Solution: We must divide $9\frac{1}{3}$ into six equal parts. Thus, each can will have

$$9\frac{1}{3} \div 6 = 9\frac{1}{3} \div \frac{6}{1} = \frac{28}{3} \times \frac{1}{6} = \frac{28}{18} = \frac{14}{9} \text{ or } 1\frac{5}{9} \text{ pounds of nuts.}$$

1. What is the value of $\frac{1}{2} \times \frac{3}{8}$?   *Answer:* _____

2. What is the value of $\frac{4}{5} \times \frac{3}{10}$?   *Answer:* _____

3. What is the value of $2\frac{2}{7} \times \frac{14}{3}$?   *Answer:* _____

4. What is the value of $\frac{9}{10} \div \frac{2}{3}$?   *Answer:* _____

5. What is the value of $\frac{8}{5} \div \frac{6}{25}$?   *Answer:* _____

6. What is the value of $\frac{4}{7} \div 3\frac{3}{11}$?   *Answer:* _____

7. What is the value of $\frac{4}{5} \div \frac{1}{6} \div \frac{2}{15}$?   *Answer:* _____

8. What is the value of $1\frac{3}{8} \div \frac{1}{4} \times \frac{2}{9}$?   *Answer:* _____

9. A zoo contains $17\frac{1}{3}$ acres of land. If $\frac{4}{13}$ of this park is used to build shelters for the animals in this zoo, how many acres of land would be used for these shelters?

    Answer: _____

10. Lorraine rode her bike a distance of $28\frac{4}{5}$ miles in 4 hours. If her speed was the same during her entire trip, how many miles did she cover in 1 hour?

    Answer: _____

LESSONS 6-9

QUIZ TWO

1. What is the ratio of 7 yards to 10 feet in reduced form?

   A $\dfrac{7}{30}$

   B $\dfrac{10}{21}$

   C $\dfrac{7}{12}$

   D $\dfrac{21}{10}$

2. What is the mixed fraction form of $\dfrac{91}{5}$?

   A $18\dfrac{1}{91}$

   B $18\dfrac{1}{5}$

   C $19\dfrac{1}{91}$

   D $19\dfrac{1}{5}$

3. What is the value of $\dfrac{4}{11} + \dfrac{3}{4}$?

   A $\dfrac{7}{44}$

   B $\dfrac{3}{11}$

   C $\dfrac{7}{15}$

   D $\dfrac{49}{44}$

4. Which one of the following fractions is already in reduced form?

   A $\dfrac{14}{21}$

   B $\dfrac{27}{36}$

   C $\dfrac{35}{39}$

   D $\dfrac{42}{45}$

5. What is the value of $7\dfrac{3}{4} - 2\dfrac{6}{7}$?

   A $5\dfrac{25}{28}$

   B $5\dfrac{3}{28}$

   C $4\dfrac{25}{28}$

   D $4\dfrac{3}{28}$

6. What is the value of $\dfrac{20}{3} - \dfrac{7}{4} + \dfrac{5}{2}$?

   A $\dfrac{89}{12}$

   B $\dfrac{49}{12}$

   C $\dfrac{19}{6}$

   D $\dfrac{8}{3}$

7. What is the value of $\dfrac{7}{15} \times \dfrac{10}{13}$?

   A $\dfrac{150}{91}$

   B $\dfrac{17}{28}$

   C $\dfrac{91}{150}$

   D $\dfrac{14}{39}$

8. What is the value of $3\frac{3}{5} \div \frac{24}{5}$?

    A $\frac{5}{21}$

    B $\frac{3}{4}$

    C $\frac{4}{3}$

    D $\frac{42}{25}$

9. What is the value of $\frac{9}{5} \div \frac{1}{3} \times \frac{8}{33}$?

    A $\frac{72}{55}$

    B $\frac{51}{40}$

    C $\frac{11}{40}$

    D $\frac{8}{55}$

10. In a shipment of radios, $\frac{5}{6}$ of them have both AM and FM frequencies. Of these dual frequency radios, $\frac{1}{8}$ of them are broken. What fraction of the shipment can be used to represent broken radios with both AM and FM?

    A $\frac{23}{24}$

    B $\frac{5}{6}$

    C $\frac{17}{24}$

    D $\frac{5}{48}$

10

Converting Percents to Fractions and Decimals

In this lesson, we will explore how to change a percent to a fraction or to a decimal. You have often seen percents used in things like (a) the percent tip that a waitress receives, (b) the percent commission that a car salesperson receives in selling a car, (c) the percent shooting average for a basketball player, (d) and the percent a student gets on an exam in math.

Your Goal: When you have completed this lesson, you should be able to change any percent number to its equivalent value as a fraction or a decimal. It is important to know that a percent shows how a number compares to 100.

LESSON 10

Converting Percents to Fractions and Decimals

Percent means hundredths. A given percent means a specific number of parts out of 100. As an example, 71% means 71 parts out of 100. As a fraction, we place the number in front of the percent sign over 100. So, 71% can be written as $\frac{71}{100}$. **To change a percent to a decimal**, the decimal point in the number will be moved two places to the left. Thus, 71%, which is really 71.%, becomes .71 as a decimal.

Example: *If 21% means 21 parts out of 100 parts, how would you write the fraction form and the decimal equivalent?*

1

Solution: We can write the ratio as $\frac{21}{100}$. When converting 21% to a decimal, we identify the (invisible) decimal point to the right of the digit 1. Thus, the decimal equivalent becomes .21, which is found by moving the decimal point two places to the left.

Example: *If 15% means 15 parts out of 100 parts, how would you write the fraction form and the decimal equivalent?*

2

Solution: We can write the ratio as $\frac{15}{100}$. However, we want the fraction to be reduced to its lowest form. This can be done by dividing both the numerator and denominator by 5, so that the fraction becomes $\frac{3}{20}$. The decimal equivalent is .15, which is found by moving the (invisible) decimal point two places to the left. Remember: Any integer has an invisible decimal point after the units digit. You won't see it, but it's always there!

Example:

3

> *If 9% means 9 parts out of 100 parts, how would you write the fraction form and the decimal equivalent?*

Solution:

> We can write the ratio as $\frac{9}{100}$. There is no further reduction of this fraction, since no number (except 1) can divide into both 9 and 100. The decimal equivalent is 0.09, which is found by moving the (invisible) decimal point two places to the left. Notice that a zero must be included to the right of the decimal point as part of the answer. The zero to the left of the decimal point is included for clarity.

Example:

4

> *If .3%, which can also appear as 0.3%, means .3 parts out of 100 parts, how would you write the fraction form and the decimal equivalent?*

Solution:

> The initial ratio is $\frac{.3}{100}$. In order to write this fraction in reduced form, we multiply the numerator and denominator by 10 to get $\frac{3}{1000}$. Multiplying a number by 10 always moves the decimal point one place to the right. The decimal equivalent is found by starting with .3 and moving the decimal point two places to the left. Then .003 is the result we get. Note that we add zero(s) if necessary.

Example:

5

> *If .016% means .016 parts out of 100 parts, how would you write the fraction form and the decimal equivalent?*

Solution:

> The initial ratio is $\frac{.016}{100}$. In order to reduce this to lowest terms, first multiply the numerator and denominator by 1000. Multiplying a number by 1000 always moves the decimal point three places to the right. The fraction will then be $\frac{16}{100,000}$. Now, to reduce the fraction to lowest terms, divide the numerator and denominator by 16 to get $\frac{1}{6250}$. By moving the decimal point two places to the left, .00016 becomes the decimal equivalent.

6

Example: *If 133% means 133 parts out of 100 parts, how would you write the fraction form and the decimal equivalent?*

Solution: We write the ratio as $\frac{133}{100}$. In fraction form, this number is considered reduced since there is no common number that can divide into both the numerator and denominator. The "invisible" decimal point lies to the right of the 3 in the units place, which means that the number really means 133.%. So by moving the decimal point two places to the left, 1.33 becomes the decimal equivalent.

7

Example: *If 2520% means 2520 parts out of 100 parts, how would you write the fraction form and the decimal equivalent?*

Solution: We write the ratio as $\frac{2520}{100}$. Now, we reduce this fraction to $\frac{126}{5}$ by dividing the numerator and denominator by 20. To change to a decimal, we move the "invisible" decimal point that lies to the right of the "0" digit two places to the left. Then, 25.20 becomes the decimal equivalent. You could also write 25.2 as the answer.

8

Example: *If 1.55% means 1.55 parts out of 100 parts, how would you write the fraction form and the decimal equivalent?*

Solution: We write the ratio as $\frac{1.55}{100}$. Now, multiply both the numerator and denominator by 100 to get $\frac{155}{10,000}$. Finally, divide both the top and bottom by 5 to get $\frac{31}{2000}$. To write the decimal equivalent, simply move the decimal point in 1.55% two places to the left and remove the percent sign. Note that a zero must be added so that 0.0155 is the decimal equivalent.

Example: *If* $8\frac{1}{3}$*% means* $8\frac{1}{3}$ *parts out of 100 parts, how would you write the*
9 *fraction form and the decimal equivalent?*

Solution: We write the ratio as $\dfrac{8\frac{1}{3}}{100}$. We then rewrite the numerator as

the improper fraction $\dfrac{25}{3}$, which will be divided by 100. Now

$\dfrac{25}{3} \div 100 = \dfrac{25}{3} \times \dfrac{1}{100} = \dfrac{25}{300} = \dfrac{1}{12}$, which is the fractional
equivalent, in reduced form.

The easiest way to change $8\frac{1}{3}$% into its decimal equivalent is
$\dfrac{1}{3} = 0.\overline{3}$. So, $8\frac{1}{3}$% = $8.\overline{3}$% = $0.08\overline{3}$ is the correct decimal equivalent.

Example: *If* $\dfrac{3}{16}$*% means* $\dfrac{3}{16}$ *parts out of 100 parts, how would you write the*
10 *fraction form and the decimal equivalent?*

Solution: The initial ratio is $\dfrac{\frac{3}{16}}{100}$. Following the exact same steps as in

Example 9, $\dfrac{\frac{3}{16}}{100} = \dfrac{3}{16} \times \dfrac{1}{100} = \dfrac{3}{1600}$, which is the fractional
equivalent.

To find the decimal equivalent, we convert $\dfrac{3}{16}$ into 0.1875 by

simply dividing 16 into 3. Finally, 0.1875% = 0.001875 by moving
the decimal point two places to the left.

MathFlash!

For an example such as 66%, be sure you don't write its fractional
equivalent as $\dfrac{2}{3}$*. In fact, 66% =* $\dfrac{66}{100}$ *=* $\dfrac{33}{50}$ *which is not* $\dfrac{2}{3}$*. The*
equivalent for $\dfrac{2}{3}$ *is actually* $66\frac{2}{3}$*%.*

Test Yourself!

1.   What is the decimal form of 2%?   Answer: _____

2.   What is the decimal form of 19.4%?   Answer: _____

3.   What is the decimal form of 0.059%?   Answer: _____

4.   What is the decimal form of 205%?   Answer: _____

5.   What is the decimal form of $12\frac{1}{8}$%?   Answer: _____

6.   What is the decimal form of 5001%?   Answer: _____

7.   What is the reduced fraction form of 16%?   Answer: _____

8.   What is the reduced fraction form of 140%?   Answer: _____

9.   What is the reduced fraction form of 0.004%?   Answer: _____

10.   What is the reduced fraction form of 7.35%?   Answer: _____

11.   What is the reduced fraction form of 24.2%?   Answer: _____

12.   What is the reduced fraction form of $\frac{2}{11}$%?   Answer: _____

Converting Decimals to Fractions and Percents

In this lesson, we will explore how to change a decimal to a fraction or to a percent. You have seen decimals used, for example, as (a) the batting average of a baseball player, (b) the body temperature of a normal person, (c) the purchase price of an item, and (d) the concentration of water in a can of juice.

Your Goal: When you have completed this lesson, you should be able to change any decimal to its equivalent value when written as a fraction or a percent. It will be important to remember that a decimal is a way of writing numbers with greater accuracy than would be possible if we only had whole numbers.

LESSON 11

Converting Decimals to Fractions and Percents

In any decimal, the first digit to the right of the decimal point represents tenths, the next digit represents hundredths, and then come thousandths, ten-thousandths, hundred-thousandths, and so forth. As a fraction, we place the number over 10, 100, 1000, and so forth, depending on how many digits exist in the decimal. For example, 0.43 would appear as $\frac{43}{100}$.

To change a decimal to a percent, we move the decimal point two places to the right and add zeros if necessary. Thus, 0.43 is equivalent to 43%.

Example: *If 0.31 means 31 hundredths, how would you write the fractional and percent equivalents?*

1

Solution: The fractional equivalent is $\frac{31}{100}$. To convert 0.31 to a percent, we move the decimal point two places to the right. The percent equivalent is 31%.

Example: *If 0.045 means 45 thousandths, how would you write the fractional and percent equivalents?*

2

Solution: The fractional equivalent is $\frac{45}{1000}$. To reduce it to its lowest form, divide both the numerator and denominator by 5. The fraction becomes $\frac{9}{200}$. To convert 0.045 to a percent, we move the decimal point two places to the right. The percent equivalent is 4.5%.

Example:

3

If 0.0006 means 6 ten-thousandths, how would you write the fractional and percent equivalents?

Solution: The fractional equivalent is $\frac{6}{10,000}$. As in Example 2, we must reduce the fraction to lowest terms. This can be done by dividing both the numerator and denominator by 2, so that the fraction becomes $\frac{3}{5000}$. To convert 0.0006 to a percent, move the decimal point two places to the right. The percent equivalent is 0.06%.

Example:

4

If 5.7 means 5 and 7 tenths, how would you write the fractional and percent equivalents?

Solution: The fractional equivalent is $5\frac{7}{10}$ or $\frac{57}{10}$. To convert 5.7 to a percent, we follow the same rules as in previous examples. However, when the decimal point is moved two places to the right, we must add a zero. Thus, the percent equivalent is 570%. Remember: Percent values may exceed 100.

Example:

5

How would you write the fractional and percent equivalents of 17?

Solution: A whole number such as 17 can be written in fractional form as $\frac{17}{1}$, but this is usually unnecessary. To convert 17 to a percent, we first note that the "invisible" decimal point is actually to the right of the digit 7. Moving the decimal point two places to the right, the answer is 1700%.

Example:

6

How would you write the fractional and percent equivalents of 1.164?

Solution: A number such as 1.164 means 1 and 164 thousandths, so we can write it in fraction form as $1\frac{164}{1000} = 1\frac{41}{250}$. The reduced improper fraction would be $\frac{291}{250}$. Moving the decimal point two places to the right, the answer as a percent will be 116.4%.

Example:

7

Solution:

If 13.9 means 13 and 9 tenths, how would you write the fractional and percent equivalents?

We can write it in fraction form as $13\frac{9}{10}$ or as $\frac{139}{10}$. To convert 13.9 to a percent, just move the decimal point two places to the right. You would have to add one zero. The answer is 1390%.

Example:

8

Solution:

How would you write the fractional and percent equivalents of a number such as $0.06\frac{1}{3}$?

This number is a mixture of decimal and fraction. To change this number to a complete fraction, notice that it can be read as "six and one-third hundredths." Now, we write $\frac{06\frac{1}{3}}{100} = \frac{\frac{19}{3}}{100} = \frac{19}{3} \times \frac{1}{100} = \frac{19}{300}$. To convert $.06\frac{1}{3}$ to a percent, we move the decimal point two places to the right to get $6\frac{1}{3}\%$, which is usually written as $6\overline{3}\%\left(\frac{1}{3} = 0.\overline{3}\right)$.

Notice how the decimal point in the number $6\frac{1}{3}\%$ is dropped. We don't write $6.\frac{1}{3}\%$ when the fraction immediately follows where the decimal point would be located.

Example:

9

Solution:

Let's try another "mixture" of decimal and fraction. How would you write the fractional and percent equivalents of the number $0.0005\frac{3}{4}$?

To change to a complete fraction, read this number as "five and three-fourths ten-thousandths." Now, write $\frac{5\frac{3}{4}}{10,000} = \frac{\frac{23}{4}}{10,000} = \frac{23}{4} \times \frac{1}{10,000} = \frac{23}{40,000}$. To convert $0.0005\frac{3}{4}$ to a percent, we will move the decimal point two places to the right to get $0.05\frac{3}{4}\%$, which can also be written as 0.0575%. Remember that $\frac{3}{4} = 0.75$.

Another option is to use the fraction $\frac{23}{40,000}$, multiply by 100, and then add the percent sign. So, $\frac{23}{40,000} \times 100 = \frac{23}{400}\%$. But $\frac{23}{400} = 0.0575$, so our answer of 0.0575% is confirmed.

You could also leave the answer as $0.05\frac{3}{4}\%$.

MathFlash!

With percents containing fractions, the fraction should be reduced. Thus, if your final answer is $\frac{8}{10}\%$, be sure to reduce it to $\frac{4}{5}\%$. Likewise, an answer of $35\frac{6}{9}\%$ should be written as $35\frac{2}{3}\%$.

Example: *How would you write the fractional and percent equivalents of the number $0.03\frac{1}{5}$?*

10

Solution: To change this number to a fraction note that it reads as $3\frac{1}{5}$ hundredths, which means $\frac{3\frac{1}{5}}{100} = \frac{\frac{16}{5}}{100} = \frac{16}{5} \times \frac{1}{100} = \frac{16}{500}$. This reduces to $\frac{4}{125}$. To convert $0.03\frac{1}{5}$ to a percent, move the decimal point two places to the right to get $3\frac{1}{5}\%$.

Since $\frac{1}{5} = 0.2$, your final answer can also be 3.2%.

Test Yourself!

1.   How is 0.275 written as a percent?   *Answer:* _____

2.   How is 8.7 written as a percent?   *Answer:* _____

3.   How is 43.21 written as a percent?   *Answer:* _____

4.   How is 0.000567 written as a percent?   *Answer:* _____

5.   How is $0.04\frac{1}{8}$ written as a percent, using no fraction in the answer?

     Answer: _____

6.   What is the reduced fraction form of 0.008?

     Answer: _____

7.   What is the reduced fraction form of .0026?

     Answer: _____

8.   What is the reduced fraction form of 1.48?

     Answer: _____

9.   What is the reduced fraction form of $0.002\frac{3}{8}$?

     Answer: _____

10. What is the reduced fraction form of $0.010\frac{1}{12}$?

     Answer: _____

Converting Fractions to Decimals and Percents

In this lesson, we will explore how to change a fraction to a decimal or to a percent. You have seen fractions in, for example, (a) the weight in ounces of a can of olives, (b) the reaction time in seconds for applying the brakes of a car, (c) the amount of cups of flour required for baking a cake, and (d) the average number of minutes to drive to work.

Your Goal: When you have completed this lesson, you should be able to change any fraction to its equivalent value as a decimal or a percent. It will be important to remember that there are two main ingredients to a fraction, namely the numerator (top) and the denominator (bottom).

LESSON 12

Converting Fractions to Decimals and Percents

A **fraction** is a ratio comparing two numbers, the numerator (top) and the denominator (bottom). For example, in the fraction $\frac{1}{3}$, 1 is being compared to 3. In the fraction $\frac{10}{4}$, 10 is being compared to 4. An integer such as 5 can also be written in fraction form, namely, $\frac{5}{1}$, in which 5 is being compared to 1.

Example: *If $\frac{1}{2}$ represents the comparison of 1 to 2, how would you write the decimal and percent equivalents?*

Solution: Its decimal equivalent, 0.5, is found by dividing 1 by 2. To convert $\frac{1}{2}$ to a percent, multiply $\frac{1}{2}$ by 100 and add the percent sign. The result is $\left(\frac{1}{2}\right)(100)\% = 50\%$.

Note that we could have used the decimal equivalent of .5 and applied the rule for changing a decimal to a percent. Do you remember that rule? In order **to change a decimal to a percent, move the decimal point two places to the right, added zero(s) if necessary, and then attach the percent sign.** If you look at .5 and move the decimal point two places to the right, adding one zero and the percent sign, you will get 50%.

Example: If $\dfrac{3}{8}$ represents the comparison of 3 to 8, how would you write the decimal and percent equivalents?

2

Solution: Its decimal equivalent, 0.375, is found by dividing 3 by 8.

$$
\begin{array}{r}
0.375 \\
8\,\overline{)\,3.000} \\
\underline{-24} \\
60 \\
\underline{-56} \\
40 \\
\underline{-40}
\end{array}
$$

To convert the fraction to a percent, multiply $\dfrac{3}{8}$ by 100 and add the percent sign. The result is $\left(\dfrac{3}{8}\right)(100)\% = 37.5\%$.

Example: If $\dfrac{7}{99}$ represents the comparison of 7 to 99, how would you write the decimal and percent equivalents?

3

Solution: Its decimal equivalent, found by dividing 7 by 99 will appear as 0.07070707..., which shows a repeating pattern. In examples such as this, place a bar over the repeating digits. Then, $0.\overline{07}$ is the way in which the decimal appears. The equivalent percent is found by the following: $\left(\dfrac{7}{99}\right)(100)\% = 7.\overline{07}\%$. Notice that the bar is used in this form. Do you see why? The reason is that the bar appeared in the decimal, so when we move the decimal point, we have to keep the bar.

Example: If $\dfrac{11}{90}$ represents the comparison of 11 to 90, how would you write the decimal and percent equivalents?

4

Solution: Its decimal equivalent, $0.1222222 = 0.1\overline{2}$, is found by dividing 11 by 90. Remember, 90 goes on the "outside" of the division symbol, and 11.00 goes on the "inside" $\left(90\,\overline{)\,11}\,\right)$. Since the only repeating digit is 2, the bar extends over the 2, not over the 1. The equivalent percent becomes $\left(\dfrac{11}{90}\right)(100)\% = 12.\overline{2}\%$. Once again, both the decimal representation and the percent show the bar symbol.

Example: *If $\frac{12}{5}$ represents the comparison of 12 to 5, how would you write the decimal and percent equivalents?*

5

Solution: This problem looks different because the numerator is larger than the denominator. Don't panic. We'll use the same approach as we did in Examples 1 and 2. Its decimal equivalent, 2.4, is found by dividing 12 by 5. Similar to Examples 1 and 2, the percent is found as follows: $\left(\frac{12}{5}\right)(100)\% = 240\%$. Note that a fraction can represent a percent exceeding 100.

Example: *Let's try another similar example in which the answer exceeds 100%. What is the decimal and percent equivalents of $3\frac{8}{11}$?*

6

Solution: Leaving the 3 alone momentarily, we can determine the decimal equivalent for $\frac{8}{11}$ by dividing 8 by 11 to get 0.727272…, which is actually written as $0.\overline{72}$. So, $3.\overline{72}$ is the decimal equivalent. The quickest way to change $3\frac{8}{11}$ to a percent is to use the decimal equivalent $3.\overline{72}$. Then simply move the decimal point two places to the right to get $372.\overline{72}\%$.

Example: *At the other extreme, some examples represent percents less than 1%. What are the decimal and percent equivalents of $\frac{4}{625}$?*

7

Solution: By dividing 4.0000 by 625, we get 0.0064 as the decimal equivalent. The percent equivalent is easily obtained by moving the decimal point to the right, so our answer is 0.64%.

Example: *As a final example, what is the decimal and percent equivalents of $\frac{31}{90,000}$?*

8

Solution: Upon dividing 90,000 into 31.0000, we get $0.0003\overline{4}$ as the decimal equivalent. Then $0.03\overline{4}\%$ becomes the percent equivalent.

Use bar notation for decimals or percents that repeat.

1. What is the decimal form of $\dfrac{7}{11}$?   *Answer:* _____

2. What is the decimal form of $\dfrac{13}{8}$?   *Answer:* _____

3. What is the decimal form of $\dfrac{5}{12}$?   *Answer:* _____

4. What is the decimal form of $\dfrac{3}{800}$?   *Answer:* _____

5. What is the decimal form of $\dfrac{6}{111}$?   *Answer:* _____

6. What is the percent form of $\dfrac{11}{5}$?   *Answer:* _____

7. What is the percent form of $\dfrac{9}{16}$?   *Answer:* _____

8. What is the percent form of $\dfrac{25}{3300}$?   *Answer:* _____

9. What is the percent form of $\dfrac{17}{90}$?   *Answer:* _____

10. What is the percent form of $\dfrac{59}{15}$?   *Answer:* _____

13

Percent Increase and Decrease

In this lesson, we will explore applications of percents in real-life situations. Specifically, you will learn about percent increases and decreases. There are many examples in everyday life that use percent changes, such as (a) the sales tax on the price of a meal in a diner, (b) the commission for a person who sells a house, (c) the drop in earnings for a company during difficult times, and (d) the amount of weight a person loses as a result of exercise.

Your Goal: When you have completed this lesson, you should be able to figure out a percent increase or decrease for any pair of numbers. You will also be able to determine a new value, if you are given a specific percent increase or decrease.

Percent Increase and Decrease

Example: *What is the percent increase from 8 to 12?*

1

Solution: The first step is to subtract 8 from 12 to get 4. A percent increase is represented initially as a fraction using the actual increase divided by the lower of the two given numbers. Thus, we have $\frac{4}{8}$. Since we need the percent increase, we calculate $\frac{4}{8} \times 100\% = 50\%$. Note that it was not necessary to reduce $\frac{4}{8}$ to $\frac{1}{2}$.

Example: *What is the percent increase from 22 to 29?*

2

Solution: First subtract 22 from 29 to get 7. The fraction to represent the increase is $\frac{7}{22}$. Then the percent increase is $\frac{7}{22} \times 100\% = 31.82\%$.

(For this lesson, all percent answers will be written to two decimal places, rounded off if necessary.)

Example: *What is the percent increase from 11.5 to 14.49?*

3

Solution: Although we are dealing with decimals, the procedure stays the same: $14.49 - 11.5 = 2.99$. Then we calculate $\frac{2.99}{11.5} \times 100\% = 26\%$.

Example: *What is the percent increase from 16 to 37.6?*

4

Solution: The first step is to calculate $37.6 - 16 = 21.6$. Then $\frac{21.6}{16} \times 100\% = 135\%$.

MathFlash!

Percent increases can be greater than 100. Remember to use the difference between the numbers as the numerator, and the lower of the two numbers as the denominator for the fraction that represents this increase.

Example: *The price of a sofa increased from $1200 to $1500. What was the percent increase?*

5

Solution: We can ignore the dollar signs, so the first step is 1500 − 1200 = 300. Then $\frac{300}{1200} \times 100\% = 25\%$.

Example: *What is the percent decrease from 20 to 11?*

6

Solution: Whenever there is a percent decrease, we need to subtract the lower number from the higher number: 20 − 11 = 9. A percent decrease is written as the ratio (fraction) of this difference divided by the higher number, which gives us $\frac{9}{20}$.

The last step is to change the fraction to a percent. We get $\frac{9}{20} \times 100\% = 45\%$.

Example: *What is the percent decrease from 64 to 53.76?*

7

Solution: The first step is to find 64 − 53.76 = 10.24. Then we multiply: $\frac{10.24}{64} \times 100\% = 16\%$. Just as with percent increase, we do not need to worry that the fraction $\frac{10.24}{64}$ is not reduced to lowest terms.

Example: *What is the percent decrease from 34.6 to 32.8?*

8

Solution: First determine 34.6 − 32.8 = 1.8. Then $\frac{1.8}{34.6} \times 100\% = 5.20\%$ is the answer. Again, note that we rounded off the percent to the nearest two places.

Example:
9

During one week, the price of gasoline dropped from $2.80 per gallon to $2.68 per gallon. What was the percent decrease?

Solution: We calculate 2.80 − 2.68 = 0.12. Then $\frac{0.12}{2.80} \times 100\% = 4.29\%$.

Example:
10

One of David's stocks dropped from $9.00 per share to $5.76 per share. What percent drop does this represent?

Solution: We first find that the drop in price is $9.00 − $5.76 = $3.24.

Ignoring the dollar signs, the percent drop is $\frac{3.24}{9.00} \times 100\% = 36\%$.

MathFlash!

When calculating a percent decrease, be careful that you use the correct denominator (the higher number), when representing the fraction that is multiplied by 100. Also, you need to know that a percent decrease cannot exceed 100.

The next types of examples will provide us with a given number to which a percent increase or decrease is assigned. We need to find the correct number following the increase or decrease. If necessary, we will round off answers to the nearest hundredth.

Example:
11

The number 54 is increased by 10%. What is the value of the new number?

Solution: The first step is to find 10% of 54. To do this, we multiply 10% by 54. Change 10% to 0.10 and multiply (0.10)(54) = 5.4. Since we are dealing with an increase, the answer will be 54 + 5.4 = 59.4.

Example:
12

The number 80 is increased by 15%. What is the value of the new number?

Solution: After changing 15% to 0.15, we calculate (0.15)(80) = 12. Since an increase is involved, the answer will be 80 + 12 = 92.

Example: *The number 7.2 is increased by 6.5%. What is the value of the new number?*

13

Solution: Both given numbers are in decimal form, but the procedure is the same. After changing 6.5% to 0.065, we calculate $(0.065)(7.2) = 0.468$. The answer will be $7.2 + 0.468 = 7.668$, which can be rounded off to 7.67.

Example: *The number 325 is increased by 145%. What is the value of the new number?*

14

Solution: Recall the *Math Flash* following Example 4. Percent increases can exceed 100. First change 145% to 1.45. Then multiply $(1.45)(325) = 471.25$. The answer becomes $325 + 471.25 = 796.25$.

MathFlash!

You may be curious as to whether there is a slightly faster way to determine the answer. Well, you would be right! An increase of 15% can be calculated by simply multiplying the given number by 1.15; an increase of 145% can be calculated by simply multiplying the given number by 2.45. Can you spot the pattern? What you do is add 100% to the given percent increase, then multiply this new percent by the given number. (Try this on any example!)

Example: *Last year, the price of a new Saturn car was $16,000. This year, the price has increased by 23.4%. What is the price of a Saturn car this year?*

15

Solution: Let's use the technique in the *Math Flash* above. Thus, instead of multiplying $16,000 by 0.234 and adding this product to $16,000, we will simply calculate 123.4% of $16,000. The answer is found by the following calculation: $(1.234)(\$16,000) = \$19,744$.

Example:

16

The number 64 is decreased by 30%. What is the value of the new number?

Solution:

The first step is (0.30)(64) = 19.2, which represents the amount of the decrease. Since a decrease is involved, the answer is 64 – 19.2 = 44.8. Notice that a decrease requires a final step of subtraction, not addition.

Example:

17

The number 112.4 is decreased by 18%. What is the value of the new number?

Solution:

After calculating (0.18)(112.4) = 20.232, we subtract this number from 112.4 to get 92.168. As in Example 13, we can round off the final answer to 92.17.

Example:

18

The number 6000 is decreased by 0.5%. What is the value of the new number?

Solution:

Be aware that 0.5% does not equal 0.5! Change 0.5% to 0.005, and then calculate (0.005)(6000) = 30. So, the answer is 6000 – 30 = 5970.

Example:

19

At Janelle's place of work, the staff of 550 people is being reduced by 8% due to reorganization. What will be the number of people left after this reduction?

Solution:

First, calculate (0.08)(550) = 44. Then, since a reduction is involved, the answer is 550 – 44 = 506.

MathFlash!

Look back at the Math Flash following Example 14. Are you wondering if a one-step approach could be used in "percent decrease" problems? The technique for this is to subtract the percent decrease from 100 and then multiply this new percent by the given number. Of course, a percent decrease could never exceed 100!

Example: *The enrollment at Central High School has dropped by 12.5% from last year to this year. Last year's enrollment was 960. What is this year's enrollment?*

Solution: Let's use the information in the *Math Flash* just mentioned. 100% – 12.5% = 87.5%. Then after changing 87.5% to 0.875, the answer is (0.875)(960) = 840.

 Test Yourself!

If necessary, you may round off your final answer to two decimal places.

1. What number is the result of 88 increased by 25%?

   Answer: _____

2. What number is the result of 175 increased by 9.2%?

   Answer: _____

3. What number is the result of 10.8 increased by 3.4%?

   Answer: _____

4. What number is the result of 6.4 increased by 160%?

   Answer: _____

5. What number is the result of 36 decreased by 19%?

   Answer: _____

6. What number is the result of 260 decreased by 70%?

   Answer: _____

 (continued)

7. What number is the result of 904.5 decreased by 0.4%?

    Answer: _____

8. What number is the result of 52 decreased by 7.3%?

    Answer: _____

9. In a local library, there are 140 math books. Due to expansion, the library's collection of math books will increase by 35%. How many math books will the library have after the expansion?

    Answer: _____

10. A pair of shoes that usually sells for $106 is now being discounted by 9%. What is the discounted price of the shoes?

    Answer: _____

Comparing Sizes of Numbers in Decimal/Fraction Form

In this lesson, we will explore how to compare the sizes of any numbers in fraction or decimal form. Some examples will contain both fractions and decimals. If it appears that we can easily find a common denominator for the given fractions, then that is a good approach to use. A sure-fire method that will work if either it seems too "lengthy" to determine a common denominator or if we have a mixture of fractions and decimals is to change all numbers to decimals. Keep in mind that any method, properly done, will yield the correct answer.

As a practical real-life example, suppose you order $\frac{2}{3}$ of a pound of roast beef in a store.

When the clerk puts the roast beef on the scale, it registers 0.68 pound. It would certainly be useful to know that the clerk gave you more roast beef than you ordered.

Your Goal: When you have completed this lesson, you should be able to compare a group of numbers in decimal form, in fraction form, or in a mixture of both forms.

LESSON 14

Comparing Sizes of Numbers in Decimal/Fraction Form

RULE 1 Given **numbers in decimal form**, add sufficient zeros as necessary so that each decimal has the same number of placeholders. Each number can then be "read" as an integer. Whichever number "reads" largest corresponds to the largest number; likewise, whichever number "reads" smallest corresponds to the smallest number.

Example:
1

How should the numbers 0.651, 0.0832, and 0.7 be arranged in order, from smallest to largest?

Solution:

The middle decimal (0.0832) has the highest number of decimal places, which is four. So, we will add zeros to ensure that each decimal has four decimal places. Then 0.6510, 0.0832, and 0.7000 is how these numbers would appear. Now visualize these numbers as appearing without decimal points. We know that 832 is less than 6510, which is less than 7000. This is how the three numbers would read. Therefore, 0.0832, 0.651, and 0.7 is the way these numbers would read from smallest to largest.

Example:
2

How should the numbers 0.25, 0.1, 0.0307, and 0.09999 be arranged in order, from smallest to largest?

Solution:

The last decimal (0.09999) has five decimal places, so we will add enough zeros to change each of the other decimals to ones that also have five decimal places. Then 0.25000, 0.10000, 0.03070, and 0.09999 is how these numbers would appear. Mentally remove the decimal points so that you are looking at 25000, 10000, 3070, and 9999. From smallest to largest, we have 3070, 9999, 10000, and 25000. Then 0.0307, 0.09999, 0.1, and 0.25 is the correct order in which these numbers should be placed.

RULE 2 Given **numbers in fraction form**, change each fraction to an equivalent fraction using a common denominator of all the fractions. Then the order of the numerators will be the correct order of the fractions. Use this rule only if a common denominator can be easily determined. Otherwise, it is recommended that you use the decimal equivalents and Rule 1.

Example: *How should the numbers $\frac{1}{2}$, $\frac{3}{5}$, and $\frac{9}{20}$ be arranged in order, from smallest to largest?*

3

Solution: In a problem such as this, we could easily change each fraction to a common denominator of 20. The fractions would appear as $\frac{10}{20}$, $\frac{12}{20}$, and $\frac{9}{20}$. Once all the fractions contain the same denominator, their size is determined by the numerator. Since 9 < 10 < 12, the correct order becomes $\frac{9}{20}$, $\frac{10}{20}$, $\frac{12}{20}$, or $\frac{9}{20}$, $\frac{1}{2}$, $\frac{3}{5}$.

Note that in the previous example, we could have changed each fraction to its equivalent decimal form. In that case, we would have 0.5, 0.6, and 0.45 as the three decimals. Using Rule 1, we would have arrived at the same order of the fractions.

Example: *How should the numbers $\frac{3}{4}$, $\frac{7}{10}$, $\frac{13}{17}$, and $\frac{19}{27}$ be arranged in order, from smallest to largest?*

4

Solution: This problem would become huge—monstrous!—if we looked for a common denominator. Let's just change each fraction to its decimal equivalent. Then 0.75, 0.7, 0.7647..., and 0.7037037... is how these decimals would appear. In this situation, you may wonder how many decimal places are really needed since two of these continue indefinitely. The good news is that you really only need enough decimal places to resolve the relative size of each number. Let's just use four decimal places. Then 0.7500, 0.7000, 0.7647, and 0.7037 is how the decimals would appear when the fractions are converted. Using the inequality symbol and removing the decimal points, 7000 < 7037 < 7500 < 7647. Thus, the correct order of the fractions is $\frac{7}{10}$, $\frac{19}{27}$, $\frac{3}{4}$, $\frac{13}{17}$.

Example: *What is an example of a decimal number with three decimal places that is larger than 0.556 and smaller than 0.567?*

5

Solution: The answer must begin either 0.55... or 0.56... If you choose the first option, then the thousandths place must exceed 6. An answer such as 0.558 would be correct. If you choose the second option, then the thousandths place must be less than 7. An answer such as 0.564 would be correct.

6

Example: *How should the numbers $\frac{8}{15}$, 0.53, $\frac{11}{21}$, 0.541, and $\frac{21}{41}$ be arranged in order, from smallest to largest?*

Solution: Here again, we could find a common denominator for the three given fractions, but we would also need to convert the decimals to fractions with the same equivalent common denominator! Let's simply change each of the fractions to decimals. Then 0.5333..., 0.53, 0.5238, 0.541, 0.5121 would be the way these numbers would first appear. We could even use just the first three decimal places before dropping the decimal points to determine the correct order. We then have 533, 530, 523, 541, and 512. Don't worry about correct rounding procedures. Now reordering, we get 512, 523, 530, 533, and 541. This translates to $\frac{21}{41}$, $\frac{11}{21}$, 0.53, $\frac{8}{15}$, 0.541 as the correct order from smallest to largest.

7

Example: *How should the numbers $\frac{7}{9}$, 0.77, $0.\overline{76}$, $\frac{18}{23}$, and $\frac{29}{37}$ be arranged in order, from smallest to largest?*

Solution: As you look at these numbers, you can probably conclude that our best approach is to change the fractions to decimals. You will also see how really close these numbers are to each other! Let's forge ahead so that $\frac{7}{9} = 0.\overline{7}$, 0.77 = 0.770, $0..\overline{76} = 0.767676...$, $\frac{18}{23} \approx$ 0.7826..., and $\frac{29}{37} = 0.\overline{783}$. It appears that we need to use three decimal places to determine the correct order for these numbers. In that case, we would have 0.777, 0.770, 0.767, 0.782, and 0.783.

The correct order of decimals should be 0.767, 0.770, 0.777, 0.782, and 0.783. Now we know that the correct order of the original numbers becomes $0.\overline{76}$, 0.77, $\frac{7}{9}$, $\frac{18}{23}$, and $\frac{29}{37}$.

Test Yourself!

For questions 1–8, arrange in order from smallest to largest.

1.   0.7889, 0.7989, 0.0885, 0.09   *Answer:* _____

2.   0.56, 0.6, 0.07, 0.108   *Answer:* _____

3.   0.1524, 0.1452, 0.42, 0.3, 0.251   *Answer:* _____

4. $\dfrac{2}{3}, \dfrac{5}{12}, \dfrac{7}{16}, \dfrac{16}{25}$ *Answer:* _____

5. $\dfrac{1}{2}, \dfrac{4}{9}, \dfrac{10}{23}, \dfrac{16}{33}, \dfrac{13}{24}$ *Answer:* _____

6. $\dfrac{3}{4}$, 0.72, 0.6, $\dfrac{11}{18}, \dfrac{5}{7}$ *Answer:* _____

7.   0.127, $\dfrac{1}{9}$, 0.1288, $\dfrac{3}{28}, \dfrac{2}{17}$ *Answer:* _____

8.   0.0954, 0.154, $\dfrac{39}{250}, \dfrac{4}{27}$, 0.149   *Answer:* _____

9.   Write a decimal number with three decimal places that is larger than 0.978 and smaller than 0.982.

      Answer: _____

10.   Write a decimal number with three decimal places that is larger than 0.84 and smaller than 0.8425.

      Answer: _____

QUIZ THREE

1. **What is the reduced fraction form of 0.018%?**

   A $\dfrac{9}{500}$

   B $\dfrac{9}{5000}$

   C $\dfrac{9}{50,000}$

   D $\dfrac{9}{500,000}$

2. **What is the decimal form of $8\dfrac{1}{5}$?**

   A 0.082

   B 0.085

   C 0.82

   D 0.825

3. **What is the percent form of $\dfrac{27}{16}$?**

   A 168.75%

   B 16.875%

   C 1.6875%

   D 0.16875%

4. **What is the decimal form of $\dfrac{6}{125}$?**

   A 0.0208

   B 0.048

   C 0.208

   D 0.48

5. **Which one of the following is larger than 0.361 but smaller than 0.3721?**

   A 0.355

   B 0.36

   C 0.371

   D 0.382

6. **If you arranged $\dfrac{10}{21}$, 0.469, $\dfrac{13}{27}$, 0.47, and $\dfrac{17}{37}$ in order, from smallest to largest, which of these would be in the middle?**

   A $\dfrac{13}{27}$

   B 0.469

   C $\dfrac{10}{21}$

   D 0.47

7. **What is the reduced fraction form of $0.07\dfrac{2}{3}$?**

   A $\dfrac{7}{100}$

   B $\dfrac{19}{250}$

   C $\dfrac{23}{300}$

   D $\dfrac{77}{1000}$

8. **What number is the result of 48 decreased by 22%?**

   A 10.56

   B 37.44

   C 47.78

   D 58.56

9. **What number is the result of 2000 increased by 0.75%?**

   A 2000.75

   B 2015

   C 2075

   D 3500

10. **Marianne is about to receive a pay increase of 14.4%. If she currently earns $9.00 per hour, what will be her hourly rate after the increase (to the nearest cent)?**

    A $9.15

    B $9.85

    C $10.15

    D $10.30

Exponents

In this lesson, we will explore the meaning of exponents. Exponents can be positive, negative, zero, or fractional, but we will only be concerned with exponents that are positive integers or zero. The advantage of using such exponents is to enable us to write large numbers without using a great many zeros. Imagine if we wanted to express the number 5 multiplied by itself 20 times.

$5 \times 5 \times 5 \times 5 \times 5 \times 5 \times 5 \times 5 \times 5 \times 5 \times 5 \times 5 \times 5 \times 5 \times 5 \times 5 \times 5 \times 5 \times 5 \times 5$

A practical example of the use of exponents with a decimal would be the amount of money accumulated when the interest is compounded every year.

Your Goal: When you have completed this lesson, you should be able to write the value of any number containing an exponent.

Exponents

An exponent attached to a specific number reveals how many times that specific number is to be multiplied by itself. Bear in mind that we are confining our discussion to exponents such as 0, 1, 2, 3, and so forth.

1

Example: *What is the value of 6^2?*

Solution: We can write 6^2 as $(6)(6)$, so its value is 36.

2

Example: *What is the value of $(-2)^3$?*

Solution: We can write $(-2)^3$ as $(-2)(-2)(-2) = -8$. Notice that the answer is negative because there are an odd number of negative signs.

3

Example: *What is the value of $\left(\dfrac{1}{2}\right)^4$?*

Solution: We can write $\left(\dfrac{1}{2}\right)\left(\dfrac{1}{2}\right)\left(\dfrac{1}{2}\right)\left(\dfrac{1}{2}\right) = \dfrac{1}{16}$.

4

Example: *What is the value of $\left(-\dfrac{3}{2}\right)^4$?*

Solution: We can write $\left(-\dfrac{3}{2}\right)\left(-\dfrac{3}{2}\right)\left(-\dfrac{3}{2}\right)\left(-\dfrac{3}{2}\right) = \dfrac{81}{16}$. Notice that the answer is positive because there are an even number of negative signs.

5

Example: *What is the value of 3^{20}?*

Solution: We would have to write the number 3 multiplied by itself 20 times. This would be quite time consuming. Fortunately, today's calculators can provide us with the answer of 3,486,784,401 in about two seconds.

Example: 6

Solution:

What is the value of (–5)⁹?

We would have to write the number –5 multiplied by itself 9 times. Granted that doing this computation would not take as much time as the number in Example 5; using a calculator would provide us with the answer of –1,953,125 in very little time.

(Be sure you understand why the answer is negative.)

Example: 7

Solution:

What is the value of (325)¹?

We recognize that the number 325 is only to be considered once. The answer is 325. Thus, any number raised to the first power (exponent of 1) is itself. This statement applies to negative numbers and to any fractions as well.

Thus, $(-4321)^1 = -4321$, and $\left(\dfrac{9}{13}\right)^1 = \dfrac{9}{13}$.

Example: 8

Solution:

What is the value of 4⁰?

It may surprise you that the answer is not 0 but 1. Some textbooks will simply define any nonzero number raised to the zero exponent to be equal to 1. There really is a logical explanation. Think of 4^3, which equals 64. If we divide this number by 4, we get 16, which is 4^2. Note that the exponent attached to the 4 dropped by one number. Now divide 16 by 4 to get 4, which is 4^1. Again, note that the exponent dropped by 1. It seems logical that when we divide again by 4, which means $4 \div 4 = 1$, the exponent would drop by 1. So the exponent would now be 0. This implies that $4^0 = 1$. This type of argument can be used for any nonzero number.

MathFlash!

0^0 is undefined. However if zero is raised to any positive integer, even 0^1, the value is zero. Thus, for example, $0^4 = 0$, because $0 \times 0 \times 0 \times 0 = 0$.

9

Example: *What is the value of* $\left(-\dfrac{2}{5}\right)^{5}$?

Solution: We can write $\left(-\dfrac{2}{5}\right)\left(-\dfrac{2}{5}\right)\left(-\dfrac{2}{5}\right)\left(-\dfrac{2}{5}\right)\left(-\dfrac{2}{5}\right) = -\dfrac{32}{3125}$. Notice that

this answer is negative since a negative sign appears an odd

number of times in the multiplication.

10

Example: *What is the value of* $\left(\dfrac{9}{4}\right)^{13}$?

Solution: Rather than actually multiplying this fraction by itself 13 times, the use of a calculator would be best. The answer is 37,876.75244, which may be rounded off to 37,876.75.

For extremely large numbers such as 8^{40} or extremely small numbers such as $\left(\dfrac{2}{33}\right)^{50}$,

your calculator will give you an approximate value in a special format. This concept will be discussed in Lesson 16.

Test Yourself!

Write your answers in integer, reduced fraction, or decimal form (rounded off to two places for decimals).

1. How can $\left(-\dfrac{8}{3}\right)\left(-\dfrac{8}{3}\right)\left(-\dfrac{8}{3}\right)$ be written using exponents?

   Answer: _____

2. What is the value of $\left(-\dfrac{4}{3}\right)^{4}$? Answer: _____

3. What is the value of 17^{5}? Answer: _____

 *(continued)*

4.   **What is the value of (–9)⁵?**   *Answer:* _____

5.   **What is the value of (–13)²?**   *Answer:* _____

6.   **What is the value of** $\left(-\dfrac{1}{8}\right)^{7}$ **?**   *Answer:* _____

7.   **What is the value of** $\left(-\dfrac{5}{4}\right)^{2}$ **?**   *Answer:* _____

8.   **What is the value of** $\left(\dfrac{11}{3}\right)^{3}$ **?**   *Answer:* _____

9.   **What is the value of (16)⁶?**   *Answer:* _____

10. **What is the value of** $\left(\dfrac{7}{2}\right)^{4}$ **?**   *Answer:* _____

Scientific Notation

In this lesson, we will explore the use of exponents as they are used with whole numbers and decimals in a format called **scientific notation**. When numbers become extremely large or extremely small, scientific notation creates a more friendly way to understand and use them. An example of a very large number would be the distance from Earth to the sun in miles. Also, a very small number would be the width of a pencil point in inches.

Your Goal: When you have completed this lesson, you should be able to convert any whole number or decimal to scientific notation and vice versa.

Scientific Notation

When a number is written in **scientific notation**, it is shown as a number between 1 and 10, multiplied by an exponent of 10. (This exponent is referred to as a power of 10.) Using a power of 10 lets us simply move the decimal point or add zeros and adjust the decimal point. For example $10^2 = 100$, which is the number 1 followed by two zeros. Likewise, $10^5 = 100,000$, which is 1 followed by five zeros. Of course, $10^0 = 1$.

Before going any further, we will need to discuss exponents that are negative integers. For our purposes, the only number that can have an exponent that is negative will be 10. Recall that $10^3 = 1000$, $10^2 = 100$, $10^1 = 10$, and $10^0 = 1$. Look at this pattern carefully. As the exponent drops by 1, the value of the given number is divided by 10. It seems reasonable (and in fact is true!) that the value of $10^{-1} = 1 \div 10 = 0.1$. Continuing with this line of reasoning, $10^{-2} = 10^{-1} \div 10 = 0.1 \div 10 = 0.01$. Then, $10^{-3} = 0.001$, $10^{-4} = 0.0001$, $10^{-5} = 0.00001$, and so forth. In general, to evaluate 10 raised to a negative exponent, its value is given by the decimal defined as 1 preceded by one fewer number of zeros to the right of the decimal point. For example, $10^{-8} = 0.00000001$.

Example: *How is the number 75 written in scientific notation?*

Solution: First we move the decimal point between 7 and 5 so that the number reads 7.5, because 7.5 is a number between 1 and 10. Next, count the number of places that the decimal point has been moved. Since 75 = 75. , the decimal point has been moved one place to the left. Therefore, in scientific notation, the answer is 7.5×10^1.

MathFlash!

Remember that a number in scientific notation must begin with a number between 1 and 10 (with only two exceptions, to be discussed later in this lesson).

2

Example: *How is the number 684.5 written in scientific notation?*

Solution: First, we move the decimal point between 6 and 8, so that the number reads 6.845. Next, count the number of places that the decimal point has been moved. Since the decimal point has been moved two places to the left, the answer is 6.845×10^2.

3

Example: *How is the number 23,600,000 written in scientific notation?*

Solution: First, we move the decimal point so that the number reads 2.36. Next, count the number of places that the decimal point has been moved. Knowing that the "invisible" decimal point is at the far right of 23,600,000, the decimal point has been moved seven places to the left. The answer is 2.36×10^7.

MathFlash!

Examples 1, 2, or 3 can be easily checked. Let us check Example 3.
10^7 *means 10,000,000 and 2.36 × 10,000,000 = 23,600,000.*

4

Example: *How is the number 6 written in scientific notation? (This is not a trick!)*

Solution: First, note that 6 is already between 1 and 10. We do not need to move the decimal point. The scientific notation should be 6×10^0.

As a review, remember that $10^0 = 1$. In fact, any number that lies between 1 and 10 can be written in scientific notation simply by multiplying the number by 10^0.

5

Example: *How is the number 0.258 written in scientific notation?*

Solution: First, we move the decimal point so that the number reads 2.58. Next, count the number of places that the decimal point has been moved. Since the decimal point has been moved one place to the right, the answer is 2.58×10^{-1}.

MathFlash!

Scientific notation must be shown as a multiplication.

6

Example: *How is the number 0.001593 written in scientific notation?*

Solution: First, we move the decimal point so that the number reads 1.593. Next, count the number of places that the decimal point has been moved. Since it has been moved three places to the right, the answer is 1.593 × 10⁻³.

7

Example: *How is the number 0.0000008 written in scientific notation?*

Solution: First, we move the decimal point so that the number reads 8., where we are allowed to drop the decimal point. Next, count the number of places that the decimal point has been moved. Since it has been moved seven places to the right, the answer is 8 × 10⁻⁷.

In order to check Example 7, we note $10^{-7} = \frac{1}{10^7} = \frac{1}{10,000,000}$.

Then $(8)\left(\frac{1}{10,000,000}\right) = 8 \div 10,000,000 = 0.0000008$.

MathFlash!

Remember that in writing a number in scientific notation, when you move the decimal point to the left, the exponent for the base 10 is positive. Likewise, when you move the decimal point to the right, the exponent for the base 10 is negative.

8

Example: *How is the number 9.25 × 10⁵ written as an integer or as a decimal?*

Solution: Since 10⁵ = 100,000, we have (9.25)(100,000) = 925,000. Notice that we just moved the decimal point five places to the right and added the necessary zeros.

Example:

9

How is the number 5.8 × 10⁻⁶ written as an integer or as a decimal?

Solution: Since $10^{-6} = \dfrac{1}{1,000,000}$, we have $5.8 \times \dfrac{1}{1,000,000} = 0.0000058$.

Notice that we just moved the decimal point six places to the left and added the necessary zeros.

Example:

10

How is the number 2.46 × 10⁻¹⁰ written as an integer or as a decimal?

Solution: Let's save some time! We know that we simply have to move the decimal point ten places to the left. Thus, adding the necessary zeros, the answer is 0.000000000246.

Test Yourself!

1.   **What is the scientific notation for 19?**

     Answer: _____

2.   **What is the scientific notation for 0.026?**

     Answer: _____

3.   **What is the scientific notation for 34,570?**

     Answer: _____

4.   **What is the scientific notation for 0.000843?**

     Answer: _____

5. **What is the scientific notation for 9,005,111?**

   Answer: _____

6. **What is the value of 1.35×10^0?**

   Answer: _____

7. **What is the value of 4.021×10^{-8}?**

   Answer: _____

8. **What is the value of 6.8×10^7?**

   Answer: _____

9. **What is the value of 1.888×10^{-5}?**

   Answer: _____

10. **What is the value of 3.0×10^{10}?**

    Answer: _____

Order of Operations

In this lesson, we will explore the methods used to evaluate an expression in which all four basic arithmetic operations may be involved. Some examples will contain exponents. As a quick reminder, an exponent is used to show how many times a number is being multiplied by itself. For example, 7^2 means $(7)(7) = 49$, and $(-2)^3$ means $(-2)(-2)(-2) = -8$. Be careful to pay attention to the rules, because in many cases you would get a wrong answer if the rules are "broken."

Your Goal: When you have completed this lesson, you should be able to evaluate an expression involving any or all of the five arithmetic operations (addition, subtraction, multiplication, division, and exponentiation—the use of exponents).

LESSON 17

Order of Operations

When there is a sequence of arithmetic steps to be performed, the order in which they are performed is critical to getting the correct answer. For example, the mathematical expression $3 + 4 \times 5$ could have two answers, depending on whether the addition or the multiplication is done first. (As we'll find out, for this example, multiplication should be done first. The answer is 23.)

RULE 1   When you have a sequence of arithmetic steps with no parentheses, the correct order to follow is:

- operations involving exponents
- multiplication and division from left to right
- addition and subtraction from left to right

1

Example: *What is the value of $2 - 6 + 5 \times 9$?*

Solution: First, multiply 5 by 9 to get 45. Then subtract 6 from 2 to get –4. Finally, add $-4 + 45 = 41$.

2

Example: *What is the value of $(-16) \div 2 - 4^2$?*

Solution: The parentheses are used so that you can see that –16 is a negative number. Often, this is done whenever negative numbers are being multiplied, divided, or used in exponentiation. First, calculate 4^2 to get 16. Then divide –16 by 2 to get –8. Finally, $-8 - 16 = -24$.

3

Example: *What is the value of $3 + 36 \div 3 \times (-6)$?*

Solution: Similar to Example 2, the parentheses are used to separate the multiplication and minus signs. First, divide 36 by 3 to get 12. Then multiply 12 by –6 to get –72. Finally, $3 + (-72) = -69$.

4

Example: *What is the value of $4 \times (-2)^3 - 12 + (-5)(-3)$?*

Solution: Note that all parentheses are used to separate the various operational symbols. First, calculate $(-2)^3 = -8$. Then $(4)(-8) = -32$. Next, multiply -5 by -3 to get 15. Finally, $-32 - 12 + 15 = -29$.

You should note that in the last calculation, you can perform the operations in any order. For example, if you decide to combine -32 and $+15$ first, the result is -17. Then -17 added to -12 yields -29.

5

Example: *What is the value of $-24 + 18 \div 6^2 - 4 + 4 \times 10$?*

Solution: It is not your imagination—the examples are getting longer! We first calculate $6^2 = 36$. Then $18 \div 36 = \dfrac{1}{2}$. (Be careful not to write 2.) Next, $4 \times 10 = 40$. Finally, $-24 + \dfrac{1}{2} - 4 + 40 = 12\dfrac{1}{2}$.

RULE 2   When you have a sequence of arithmetic steps **with parentheses**, you complete the operations within these parentheses first and then follow the same order as listed in Rule 1.

To make our work interesting, we will return to each of the previous examples and toss in a set of parentheses. Watch carefully how the answers change.

6

Example: *What is the value of $(2 - 6 + 5) \times 9$?*

Solution: First, combine the three numbers inside the parentheses: $2 - 6 + 5 = 1$. You add 2, -6, and 5 in any order. If you decide to first combine the second and third numbers, make sure you use -6, not 6. Finally, $1 \times 9 = 9$.

7

Example: *What is the value of $(-16) \div (2 - 4)^2$?*

Solution: Because the subtraction is in the parenthesis, it must be done first: $2 - 4 = -2$. Then $(-2)^2 = 4$. (Remember that squaring a negative number always yields a positive number.) Finally, $(-16) \div 4 = -4$.

If you should come across a calculation such as –3², the squaring only applies to 3, not to –3. So, –3² = –9.

8

Example: *What is the value of (3 + 36) ÷ 3 × (–6)?*

Solution: First, add 3 to 36 to get 39. Next, 39 ÷ 3 = 13. Finally, 13 × (–6) = –78. Do not multiply before you divide! (Had you made that mistake, the final calculation would have been $39 ÷ (–18) = –2\frac{1}{6}$, which is wrong!)

9

Example: *What is the value of 4 × [(–2)³ – 12 + (–5)(–3)]?*

Solution: Note that we are using brackets to symbolize a second set of parentheses. The purpose is just for ease of reading. The first step is (–2)³ = –8. Second, within the brackets, (–5)(–3) = 15. Next, –8 – 12 + 15 = –5. Finally, 4 × (–5) = –20. Notice that the multiplication involving 4 comes last because the 4 is outside the brackets.

10

Example: *What is the value of (–24 + 18) ÷ 6² – (4 + 4 × 10)?*

Solution: This may look hard, but let's work it out. Since there are two sets of parentheses, we must do the calculations inside each one of them: –24 + 18 = –6 and 4 + 4 × 10 = 4 + 40 = 44. (Caution: Make sure you didn't mistakenly calculate 4 + 4 × 10 as 80!) Now the problem reads as follows: –6 ÷ 6² – 44. After calculating 6² as 36, we get $–6 ÷ 36 – 44 = –\frac{1}{6} – 44 = –44\frac{1}{6}$.

1.   **What is the value of $12 \div 3 + 3 \times 6$?**   *Answer:* _____

2.   **What is the value of $18 + 4^3 - 8 \times 5$?**   *Answer:* _____

3.   **What is the value of $9 - 4 \times (-5)$?**   *Answer:* _____

4.   **What is the value of $(60 - 5^2) \div 7 - 12$?**   *Answer:* _____

5.   **What is the value of $90 \div 6 - 3^3 \times 5$?**   *Answer:* _____

6.   **What is the value of $20 \times (40 - 10 \div 2 + 8)$?**   *Answer:* _____

7.   **What is the value of $(-1)^8 + (6 - 12 \div 4 \times 6)$?**   *Answer:* _____

8.   **What is the value of $-15 - 5 \times 3^3$?**   *Answer:* _____

9.   **What is the value of $28 \div (4 - 18) - 2^6$?**   *Answer:* _____

10.   **What is the value of $(-4)^3 + (-18 \div 3 + 3 \times 5)^2$?**   *Answer:* _____

LESSONS 15-17

QUIZ FOUR

1. Which one of the following has a value of 2?

    A $1^7 - (-1)^5$

    B $1^4 - (-1)^2$

    C $(2^2 - 2)^0$

    D $\left(\frac{1}{2}\right)^2 + 1^3$

2. What is the value of 4.83×10^{-4}?

    A 0.0000483

    B 0.000483

    C 0.00483

    D 0.0483

3. What is the correct scientific notation for 526,000?

    A 5260×10^2

    B 526×10^3

    C 52.6×10^4

    D 5.26×10^5

4. What is the value of $(85 - 3^4) + 20 \div (-2)$?

    A −12

    B −6

    C 14

    D 63

5. What is the value of $-8 - (-6)^2 \div 4 + (-1)^{11}$?

    A −8

    B −12

    C −16

    D −18

6. What is the correct scientific notation for 10?

    A 0.01×10^3

    B 0.1×10^2

    C 1.0×10^1

    D 10.0×10^0

7. How many zeros would be included in the number 4.007×10^{15} when it is written as an integer?

    A 16

    B 15

    C 14

    D 13

8. **What is the value of** $\left(-\dfrac{13}{3}\right)^2$?

     A $\dfrac{169}{9}$

     B $\dfrac{26}{9}$

     C $-\dfrac{26}{9}$

     D $-\dfrac{169}{9}$

9. **What is the value of** $(-2)^3 - \left(\dfrac{1}{2}\right)^2$?

     A $-7\dfrac{1}{4}$

     B $-7\dfrac{1}{2}$

     C $-8\dfrac{1}{4}$

     D $-8\dfrac{1}{2}$

10. **Which one of the following has the smallest value?**

     A 8.32×10^{-9}

     B 2.38×10^{-10}

     C 1.32×10^{-7}

     D 5.37×10^{-8}

Divisibility Rules for Integers

In this lesson, we will explore the rules that govern divisibility with "popular" numbers. These are numbers that you see most often, such as 2, 3, and 5. Without even using a calculator, we will discover easy ways to determine if a given number is divisible by one of these "popular" numbers. It is important to remember that the word "divisible" means "no remainder." As an example, you probably know that any even number is divisible by 2. Also, you could easily check that the number 57 is divisible by 3.

Now imagine that you and eight other people go to a restaurant, where your total bill is $207. Without a calculator, we'll show why a number such as this one is divisible by 9; thus, if the bill is split evenly among all nine people, each person will pay a whole number of dollars. (Hopefully, this will make sense, even without cents!).

Your Goal: When you have completed this lesson, you should know the divisibility rules for nearly every number up through 10.

LESSON 18

Divisibility Rules for Integers

A number such as 16 is divisible by 2, a number such as 70 is divisible by 10, but a number such as 55 is not divisible by 9. In this lesson, we will explore divisibility rules for integers 2, 3, 4, 5, 6, 8, 9, and 10.

RULE 1   Every number is divisible by 1.

RULE 2   A number is divisible by 2 if it ends in an even digit. The even digits are 0, 2, 4, 6, and 8.

Example: *Given the numbers 46 and 157, which is divisible by 2?*

Solution: The number 46 is divisible by 2 since it ends in an even digit, namely, 6. However, the number 157 is not divisible by 2, since it is an odd number.

RULE 3   A number is divisible by 3 if the sum of its digits is a number that is divisible by 3.

Example: *Given the numbers 84 and 2476, which is divisible by 3?*

Solution: The number 84 is divisible by 3 since when we add 8 and 4, the sum is 12, and 12 is divisible by 3. However, the number 2476 is not divisible by 3; the sum of its digits is 19, and 19 is not divisible by 3.

RULE 4   A number is divisible by 4 if the number named in the tens and units places is divisible by 4.

Example: *Given the numbers 568, 2300, and 874, which are divisible by 4?*

Solution: The number 568 is divisible by 4, since 68 is divisible by 4. The number 2300 is divisible by 4, since 00, which is actually zero, is divisible by 4. (Remember that zero divided by any nonzero number is always zero!) The number 874 is not divisible by 4, since 74 is not divisible by 4.

RULE 5   A number is divisible by 5 if it ends in 0 or 5.

Example: *Given the numbers 365, 1870, and 46,132, which are divisible by 5?*

Solution: The numbers 365 and 1870 are divisible by 5, but the number 46,132 is not divisible by 5.

RULE 6   A number is divisible by 6 if it is divisible by both 2 and by 3. Any odd number **cannot** be divisible by 6, since it could never be divisible by 2.

Example: *Given the numbers 288 and 752, which is divisible by 6?*

Solution: The number 288 is divisible by 2 since it is even and the sum of its digits, which is 18, is divisible by 3. Therefore, this number is divisible by 6. The number 752 is not divisible by 6; the sum of its digits, which is 14, is not divisible by 3.

RULE 7   Only used in higher-level math.

RULE 8   A number is divisible by 8 if the number named in the hundreds, tens, and units places is divisible by 8. Although similar to the rule for divisibility by 4, this rule is probably the least helpful, if only because we need to inspect (usually) a three-digit number.

Example: *Given the numbers 9480, 28,712, and 52,000, which are divisible by 8?*

Solution: The number 9480 is divisible by 8 since it can be fairly easy to see that 480 is divisible by 8. The number 28,712 is also divisible by 8, but it is not as easy to see this fact. To do so, we would check the number 712, which is divisible by 8. Similar to the rule for divisibility by 4, the number 52,000 is divisible by 8 since 000 is equivalent to zero. (Zero is divisible by any nonzero number.)

RULE 9 A number is divisible by 9 if the sum of its digits is a number that is divisible by 9. This rule is similar to the rule for divisibility by 3.

Example: *Given the numbers 522, 7398, and 15,045, which are divisible by 9?*

Solution: The number 522 is divisible by 9 since the sum of its digits is 9. The number 7398 is divisible by 9 since the sum of its digits is 27, and 27 is divisible by 9. The number 15,045 is not divisible by 9; the sum of its digits is 15, and 15 is not divisible by 9.

RULE 10 A number is divisible by 10 if it ends in a 0.

Example: *Given the numbers 54,790 and 6234, which is divisible by 10?*

Solution: The number 54,790 is divisible by 10, but the number 6234 is not divisible by 10.

Test Yourself!

1. Which one of the following numbers is divisible by 4?

   (A) 1046   (C) 2466

   (B) 1532   (D) 3478

2. If a number is divisible by 10, then it must also be divisible by which one of the following numbers?

   (A) 8   (C) 6

   (B) 7   (D) 5

3. Which one of the following numbers is divisible by 3 but is not divisible by 9?

   (A) 328   (C) 639

   (B) 542   (D) 777

4. Given the number 436_ , which represents a four-digit number, how many different digits could be entered into the units place so that the entire number is divisible by 8?

(A) 4   (C) 2

(B) 3   (D) 1

5. If a number is divisible by 6, which one of the following statements is correct?

(A) It is divisible by 2, but not necessarily divisible by 3.

(B) It is divisible by 3, but not necessarily divisible by 2.

(C) It must be divisible by both 2 and 3.

(D) It need not be divisible by either 2 or 3.

6. A number is known to be divisible by 5 but is <u>not</u> divisible by 2. Which one of the following statements is correct?

(A) The units digit must be 5.

(B) The number must be even.

(C) The number must be divisible by 10.

(D) Each digit of the number must be 5.

7. Consider the number 5_9_ . Which one of the following digits can be entered into both the hundred's place and the unit's place so that the entire number is divisible by 6?

(A) 0   (C) 4

(B) 2   (D) 6

Test Yourself! (continued)

8.   Which number between 10 and 50 is divisible by 2, 3, and 5?

(A)   30   (C)   15

(B)   20   (D)   12

9.   Given the number 67_1, how many different digits could be entered into the tens place so that the entire number is divisible by 3?

(A)   1   (C)   3

(B)   2   (D)   4

10.   Which one of the following numbers is divisible by 3 and 4 but is not divisible by 9?

(A)   90   (C)   44

(B)   60   (D)   36

Primes and Composites

In this lesson, we will explore the meaning of prime and composite numbers. We will only use positive integers in this lesson. You will notice that the size of a number has no relationship to whether it is prime or composite.

Your Goal: When you have completed this lesson, you should be able to identify, for most integers less than 100, which ones are prime and which ones are composite.

LESSON 19

Primes and Composites

A **prime number** is defined as a number that is divisible by exactly two numbers, itself and 1. Examples are 2, 11, 29, and 47. Another way to say "is divisible by" is "has factors of." As an example, the only two factors of 11 are 1 and 11.

A **composite number** is defined as a number that is divisible by at least three numbers. Since any number is automatically divisible by itself and by 1, a composite number must contain at least one additional factor. Examples are 6 and 65. The factors of the number 6 are 1, 2, 3, and 6. The factors of 65 are 1, 5, 13, and 65.

The list of primes, as well as the list of composites is infinite. No matter what prime number you have, there will always be a prime number that is larger. Currently, the largest known prime number (discovered in November 2003) contains 6,320,430 digits!

With composite numbers, no matter what composite number is stated, we just have to locate the next even number to guarantee a larger composite number. (Remember: Every even number is divisible by 2.)

The number 1 is neither prime nor composite. The reason is that 1 has only a single factor, namely, itself. A prime number must have exactly two factors. When you list all the factors of any number, 1 and the number itself must be included.

Example: *What are all the factors of 12?*

Solution: Start with 1 and 12 to list two factors. Next, note that since 12 is even, it must have a factor of 2; 12 ÷ 2 = 6, so 2 and 6 are factors. Continuing with 3, 12 ÷ 3 = 4, so 3 and 4 are factors. The next number to be checked would have been 4, but we have already discovered that 4 is a factor. Our list is now complete. The factors of 12 are 1, 2, 3, 4, 6, and 12.

2 **Example:** *What are all the factors of 25?*

Solution: Start with 1 and 25. We note that 2, 3, and 4 are not factors, but 5 is a factor. Since 25 ÷ 5 = 5, we are finished.
The factors of 25 are 1, 5, and 25.

3 **Example** *What are all the factors of 40?*

Solution: Start with 1 and 40. We know that 40 ÷ 2 = 20, so both 2 and 20 are factors. Next, 40 ÷ 3 is not an integer, so we skip over the number 3. However, 40 ÷ 4 = 10, so both 4 and 10 are factors. Are we done? We still need to check any number between 4 and 10. Hopefully, you have concluded that 5 is a factor of 40, and sure enough, 40 ÷ 5 = 8. This means that 5 and 8 are factors of 40. The only two numbers left to check are 6 and 7, but neither is a factor of 40. The factors of 40 are 1, 2, 4, 5, 8, 10, 20, and 40.

4 **Example:** *What are all the factors of 63?*

Solution: After recording 1 and 63 as factors, we note that 63 ÷ 3 = 21, so 3 and 21 are factors. After 3, the next number we locate that is a factor of 63 is 7, since 63 ÷ 7 = 9. Then 7 and 9 are factors, and since 8 is not a factor of 63, our list will be complete.
The factors of 63 are 1, 3, 7, 9, 21, and 63.

5 **Example:** *What are all the factors of 87?*

Solution: As usual, start with 1 and 87. Next, note that 3 is a factor of 87, and 87 ÷ 3 = 29. How can we tell if we are done? You would need to find a number between 3 and 29 that divides into 87. Using any of the rules you learned in Lesson 18, plus simply testing additional numbers, you will not find any other factors.
So, the factors of 87 are 1, 3, 29, and 87.

6 **Example:** *What are all the factors of 96?*

Solution: As usual, start with 1 and 96. Next, 96 ÷ 2 = 48, so 2 and 48 are factors. Since 96 ÷ 3 = 32, 3 and 32 are also factors. Here are the subsequent divisions to find the remaining factors: 96 ÷ 4 = 24, 96 ÷ 6 = 16, 96 ÷ 8 = 12. The factors of 96 are 1, 2, 3, 4, 6, 8, 12, 16, 24, 32, 48, and 96. Since the last two numbers found in the division process are 8 and 12, we are done, unless there is a factor between 8 and 12. You can easily check that 9, 10, and 11 do not divide into 96 exactly, so our list is complete.

7

Example: *What are all the factors of 61?*

Solution: Start with 1 and 61. Notice that none of the divisibility rules that you learned in Lesson 18 seem to be applicable. You could try numbers for which we have not specified any divisibility rule, such as 7 or 13, but these are not factors of 61.

The number 61 is prime, so it only has factors of 1 and 61.

Test Yourself!

1. How many prime numbers are smaller than 10?

   (A)   1   (C)   3

   (B)   2   (D)   4

2. Which prime number is larger than 26 but smaller than 30?

   Answer: _____

3. Write all the two-digit prime numbers that end in the digit 3.

   Answers: _____

4. Write all the factors of 39.

   Answer: _____

5. Write all the factors of 18.

   Answer: _____

Test Yourself! (continued)

6.   Which one of the following numbers is <u>not</u> a factor of 60?

    (A)   12   (C)   26

    (B)   20   (D)   30

7.   Write all the factors of 75 that are composite.

    Answer: _____

8.   How many composite numbers are there below 80 that end in the digit 7?

    (A)   1   (C)   3

    (B)   2   (D)   4

9.   Which even number is prime?

    Answer: _____

10.   Write all the factors of 52 that are prime.

    Answers: _____

Prime Factorization

In this lesson, we will explore the meaning of prime factorization of numbers. From Lesson 19, we learned that positive integers (except 1) can be categorized as either prime or composite. As in Lesson 19, we will only use positive integers. You will discover that there is a unique prime factorization for every number. Also, we will be using exponents to write the prime factorization in compact form.

Your Goal: When you have completed this lesson, you should be able to write the prime factorization of any number.

LESSON 20

Prime Factorization

If a number is already prime, such as 13, you simply write 13 to show its prime factorization. (Can you guess why we don't write 13 × 1? The answer is that 1 is not prime.) **Prime factorization** is the writing of each number as a product of only prime numbers. If a prime number is repeated any number of times, we will use exponents.

1

Example: *What is the prime factorization of the number 12?*

Solution: The first step is to find one factor of 12. We know that 12 can be divided by 2, and that (2)(6) = 12. The number 2 is prime, but 6 is not. The next step is to write 6 as a product of two numbers, namely, (2)(3). The final answer becomes (2)(2)(3), which is commonly written as $2^2 \times 3$.

MathFlash!

It will be important to remember that no matter what path you use to subdivide a number into its "parts," the final answer will always remain the same. In Example 1, suppose you used (3)(4) = 12 initially. Then $4 = 2^2$, so your answer would still be 2×3. Also, for consistency, we place the prime factors in ascending order. (This is only for convenience. 3×2^2 or even as $2 \times 3 \times 2$ would still be considered correct.)

2

Example: *What is the prime factorization of the number 39?*

Solution: Recognizing that 3 is a factor of 39, by division we can find the other factor, namely, 13. Then the final answer becomes 3 × 13.

3

Example: *What is the prime factorization of the number 80?*

Solution: We could use (2)(40), but you will find that you can arrive at the correct solution quickest if you initially use factors that are "close" to each other in value. In this example, let's use (8)(10). We know that $8 = 2^3$ and that $10 = 2 \times 5$. Putting these pieces together, we get $2^3 \times 2 \times 5 = 2^4 \times 5$.

MathFlash!

If you did use (2)(40), the process of prime factorization would simply take more steps. Here is how it might look: Since $40 = (4)(10)$, the next step would appear as (2)(4)(10). Now, rewrite 4 as (2)(2), and 10 as (2)(5). Then the prime factorization would appear as $(2)(2)(2)(2)(5) = 2^4 \times 5$.

4

Example: *What is the prime factorization of the number 145?*

Solution: It is easy to recognize 5 as a factor, so initially, we have (5)(29). So, 5×29 is the final answer.

5

Example: *What is the prime factorization of the number 539 ?*

Solution: This may look quite difficult because none of the rules you studied about divisibility seem to be useful here. But remember that you can always check "small" numbers for which we have not established any automatic divisibility rules. Can you guess what number we should try? If you are guessing that 7 or 11 would be the best choice, you are right; $539 \div 7 = 77$, and we know that $77 = (7)(11)$. The final answer is $7 \times 7 \times 11 = 7^2 \times 11$.

6

Example: *As a final example, let's try one with many steps. What is the prime factorization of the number 3744?*

Solution: You could begin this example by recognizing that 9 must be a factor since the sum of the digits of 3744 is 18, which is divisible by 9. Upon division, 3744 = (9)(416). Hopefully, you can easily recognize that 416 must be divisible by 4. Now, 416 = (4)(104). But 104 = (4)(26), and 26 = (2)(13).

Thus, we have $3744 = 9 \times 416 = 3^2 \times 4 \times 104 = 3^2 \times 2^2 \times 4 \times 26 = 3^2 \times 2^2 \times 2^2 \times 2 \times 13$. (Don't forget to write 9 as 3^2.)

Finally, we get the answer of $2^5 \times 3^2 \times 13$.

MathFlash!

You certainly could have begun the last example by identifying 4 as a factor of 3744, and you would first have (4)(936). If you continued carefully, you would end up with the answer of $2^5 \times 3^2 \times 13$.

Although we used only one way to get to the final answer for each example, you are encouraged to try a different approach for each example. By doing this, you will strengthen your understanding of prime factorization. Also, it is important to realize that you can check any of these answers by multiplying all the factors. For example, as in the *Math Flash* above, $2^5 \times 3^2 \times 13 = (32)(9)(13) = 3744$.

Test Yourself!

1.   **What is the prime factorization of 40?**

   Answer: _____

2.   **What is the prime factorization of 168?**

   Answer: _____

3.   **What is the prime factorization of 63?**

     Answer: _____

4.   **What is the prime factorization of 325?**

     Answer: _____

5.   **What is the prime factorization of 91?**

     Answer: _____

6.   **What is the prime factorization of 539?**

     Answer: _____

7.   **What is the prime factorization of 2992?**

     Answer: _____

8.   **What is the prime factorization of 621?**

     Answer: _____

9.   **What is the prime factorization of 224?**

     Answer: _____

10. **What is the prime factorization of 1178?**

     Answer: _____

Greatest Common Factor (GCF)

In this lesson, we will explore how to find the greatest common factor (or GCF) of two or three numbers. Technically, the process we will use can apply to the GCF of more than three numbers. We will use the same procedure for each example, even though at times the answer may seem obvious. In fact, your understanding of Lesson 20 is very important to learning this lesson. You will need to remember that there is a GCF for every example. If two numbers do not share a common factor, the GCF will be 1.

Your Goal: When you have completed this lesson, you should be able to write the GCF of any pair of numbers.

Greatest Common Factor (GCF)

The **GCF of two or more numbers** is the largest number that is a factor of all the given numbers. Sometimes the answer may seem obvious. For example, the GCF of 4 and 6 is 2. This would be written as GCF(4, 6) = 2. We can easily see that 2 is a factor of each, and no higher number is a factor of both. As another example, suppose we wanted the GCF of 7 and 12. The only factors of 7 are 1 and 7. Which one of these is also a factor of 12? You would be right if you said 1. Thus, GCF(7, 12) = 1. Let's do some examples that require a system to determine the GCF.

1

Example: *What is the GCF of the numbers 10 and 45?*

Solution: Rewrite each number in prime factorization form, $10 = 2 \times 5$ and $45 = 3^2 \times 5$. The method we will use is to locate only common bases and then use the lower exponent of those bases. For this example, the only common base is 5. Since each number has a 5 (no 5^2, 5^3, 5^4, etc.), 5 is the GCF.

MathFlash!

Quick review: You may recall that for a number such as 3^2, the base is 3, and the exponent is 2. For a number such as 5, we know that $5 = 5^1$, so the base is 5, and the exponent is 1.

2

Example: *What is the GCF of the numbers 88 and 36?*

Solution: Rewrite each number in prime factorization form, $88 = 2^3 \times 11$ and $36 = 2^2 \times 3^2$. Can you spot the common base? It is 2. Now look at the exponents for the base 2. The lower of the two exponents is 2. Then $2^2 = 4$ is the GCF.

3

Example: *What is the GCF of the numbers 99 and 330?*

Solution: Rewriting each number in prime factorization form, $99 = 3^2 \times 11$ and $330 = 2 \times 3 \times 5 \times 11$. Both 3 and 11 are common bases. The lower exponent for the two 3s is 1, and each exponent for 11 is 1. Then $3 \times 11 = 33$ is the GCF.

4

Example: *What is the GCF of the numbers 29 and 83?*

Solution: Hopefully, you recognize these as prime numbers. As such, the GCF is 1.

5

Example: *What is the GCF of the numbers 43 and 559?*

Solution: You may already know that 43 is a prime number, but how should you handle 559? Once you realize that none of our divisibility rules work, you just need to patiently check "relatively" small numbers. The number 13 may have a reputation as "bad luck," but not in this example! Sure enough, 13 is a factor of 559, and by division, we can see that $559 = (13)(43)$. The common base is 43, and this number is the GCF.

MathFlash!

Given any two numbers for which one of them is a factor of the other, the GCF will be the lower number. As another example, the GCF of 16 and 864 is 16, since 16 is a factor of 864.

6

Example: *What is the GCF of the numbers 735 and 350?*

Solution: Rewriting each number in prime factorization form, $735 = 3 \times 5 \times 7^2$ and $350 = 2 \times 5^2 \times 7$. Notice that both 5 and 7 are common to these numbers, so $5 \times 7 = 35$ is the GCF.

7

Example: *What is the GCF of the numbers 81 and 513?*

Solution: Proceeding as we have in all the other examples, $81 = 3^4$ and $513 = 3^3 \times 19$. The common base is 3, so all you need to identify is the lower exponent. If you answered 3, you are correct. The GCF is $3^3 = 27$.

8

Example: *What is the GCF of the numbers 18, 27, and 63?*

Solution: In prime factorization form, $18 = 2 \times 3^2$, $27 = 3^3$, and $63 = 3^2 \times 7$. The only common base is 3, and the lowest common exponent is 2. The GCF is $3^2 = 9$.

9

Example: *What is the GCF of the numbers 6, 35, and 121?*

Solution: In prime factorization form, $6 = 2 \times 3$, $35 = 5 \times 7$, and $121 = 11^2$. There is no common base, so the GCF = 1. Remember that 1 is <u>always</u> a factor of any integer.

10

Example: *What is the GCF of the numbers 196, 308, and 560?*

Solution: This will take a little extra time! In prime factorization form, $196 = 2^2 \times 7^2$, $308 = 2^2 \times 7 \times 11$, and $560 = 2^4 \times 5 \times 7$. The common bases are 2 and 7. Use the lowest common exponent for each base to see that the GCF is $2^2 \times 7 = 28$.

Test Yourself!

Write the Greatest Common Factor (GCF) for each set of numbers.

1.   30 and 42   Answer: _____

2.   160 and 90   Answer: _____

3.   135 and 297   Answer: _____

4.   91 and 156   Answer: _____

5.   25 and 51   Answer: _____

6.   120 and 345   Answer: _____

7.   41 and 656   Answer: _____

8.   280 and 2080   Answer: _____

9.   35, 65, and 60   Answer: _____

10. 270, 216, and 486   Answer: _____

Least Common Multiple (LCM)

In this lesson, we will explore how to find the least common multiple (also abbreviated as the LCM) of two or three numbers. Only positive integers will be used. There is a procedure that can be used for each example, although the answer may seem obvious at times. You must have a thorough understanding of Lessons 20 and 21 in order to learn this lesson. It will be important to remember that there is an LCM for every example.

Your Goal: When you have completed this lesson, you should be able to write the LCM of any pair of numbers. To keep things simple, we will use the same pairs of numbers that we had in Lesson 21.

Least Common Multiple (LCM)

The **LCM of two or more numbers** is the smallest number for which each of the given numbers is a factor. (Remember: *Factor* means "divisor.") There are times when the answer will be quite evident. For example, given the numbers 2 and 3, the LCM is 6. This is easy to check since 2 and 3 are factors of 6. Also, we cannot find a smaller number than 6 that has this property. Of course, there are larger numbers, such as 120, for which 2 and 3 are factors. But, we are only looking for the smallest.

Let's return to the same numbers from Lesson 21 in order to show a system we will use to determine the LCM.

1

Example: *What is the LCM of the numbers 10 and 45?*

Solution: Write the prime factorization of each number, $10 = 2 \times 5$ and $45 = 3^2 \times 5$. Identify each different base and then use the highest exponent of those bases. In this example, the bases are 2, 3, and 5. The highest exponent for the bases 2 and 5 is 1, but the highest exponent for the base of 3 is 2. Then the LCM $= 2 \times 3^2 \times 5 = 90$.

2

Example: *What is the LCM of the numbers 88 and 36?*

Solution: Write the prime factorization: $88 = 2^3 \times 11$ and $36 = 2^2 \times 3^2$. The highest exponent for the base of 2 is 3, the only exponent for the base of 3 is 2, and the only exponent for the base of 11 is 1. (Remember that when a base has an exponent of 1, the exponent is often omitted.) Then the LCM $= 2^3 \times 3^2 \times 11 = 792$.

3

Example: *What is the LCM of the numbers 99 and 330?*

Solution: $99 = 3^2 \times 11$ and $330 = 2 \times 3 \times 5 \times 11$. Use the highest exponent of 2 for the base of 3 and an exponent of 1 for the bases 2, 5, and 11. Then the LCM $= 2 \times 3^2 \times 5 \times 11 = 990$.

Example: *What is the LCM of the numbers 29 and 83?*

4

Solution: These are prime numbers. As such, each number is in its prime factorization form. Can you guess what the LCM should be? Hopefully, you concluded that the LCM is simply the multiplication (product) of these numbers. Thus, the LCM = 29 × 83 = 2407.

Example: *What is the LCM of the numbers 43 and 559?*

5

Solution: The number 43 is already in its prime factorization form, and 559 = 13 × 43. The only bases we can use are 13 and 43, and the corresponding exponent is 1 for each of them. Then the LCM = 13 × 43 = 559.

MathFlash!

Given two numbers for which one number is a factor of the other, the LCM will be the higher number. As another example, the LCM of 18 and 432 is 432, since 18 is a factor of 432.

Example: *What is the LCM of the numbers 735 and 350?*

6

Solution: $735 = 3 \times 5 \times 7^2$ and $350 = 2 \times 5^2 \times 7$. For the LCM, we will use an exponent of 1 for bases 2 and 3, but we must use an exponent of 2 for bases 5 and 7. Remember, you need the highest exponent for each base. Then the LCM = $2 \times 3 \times 5^2 \times 7^2 = 7350$.

Example: *What is the LCM of the numbers 81 and 513?*

7

Solution: $81 = 3^4$ and $513 = 3^3 \times 19$. For the LCM, we need to use the highest exponent for 3, which is 4; we also need to use 19. Then the LCM = $3^4 \times 19 = 1539$.

8

Example: *What is the LCM of the numbers 18, 27, and 63?*

Solution: $18 = 2 \times 3^2$, $27 = 3^3$, and $63 = 3^2 \times 7$. For the LCM, remember that we need the highest exponent of each base. In this example, the LCM $= 2 \times 3^3 \times 7 = 378$.

9

Example: *What is the LCM of the numbers 6, 35, and 121?*

Solution: The prime factorization for each number is as follows: $6 = 2 \times 3$, $35 = 5 \times 7$, and $121 = 11^2$. Thus, the LCM $= 2 \times 3 \times 5 \times 7 \times 11^2 = 25{,}410$. (Since no pair of these numbers shares any factor other than 1, the LCM is really just the product of the three numbers.)

10

Example: *What is the LCM of the numbers 196, 308, and 560?*

Solution: $196 = 2^2 \times 7^2$, $308 = 2^2 \times 7 \times 11$, and $560 = 2^4 \times 5 \times 7$. Thus, the LCM $= 2^4 \times 5 \times 7^2 \times 11 = 43{,}120$.

Test Yourself!

Write the LCM for each set of numbers. Save yourself some time! These are the same numbers used in Lesson 21.

1.   30 and 42   *Answer:* _____

2.   160 and 90   *Answer:* _____

3.   135 and 297   *Answer:* _____

4.   91 and 156   *Answer:* _____

5.   25 and 51   *Answer:* _____

Test Yourself! (continued)

6.   120 and 345   *Answer:* _____

7.   41 and 656   *Answer:* _____

8.   280 and 2080   *Answer:* _____

9.   35, 65, and 60   *Answer:* _____

10.   270, 216, and 486   *Answer:* _____

The Relationship between Numbers and Letters

In this lesson, we will explore some basic concepts that will be helpful when you begin your study of algebra. With all that you have learned over the previous 22 lessons, you should feel very prepared to study the branch of mathematics that deals with letters (known as variables) as well as numbers. You will discover that many concepts in mathematics can be written into algebraic formulas.

Your Goal: When you have completed this lesson, you should be able to appreciate the need for learning the concepts of algebra.

The Relationship between Numbers and Letters

In algebra, **a letter (known as a variable) is used to represent an unknown quantity**. Any letter of the alphabet can be used, but the letter x is the most common. To help identify variables, they will always be italicized. These variables can be combined with numbers and symbols in order to express a word statement. For each of the following examples, the letter x will be used to represent the unknown quantity.

Example:
1

How is "the sum of an unknown number and eight" expressed in mathematical symbols?

Solution:

We recall that "the sum of" is equivalent to the "+" symbol. The answer is written as $x + 8$. We do not know the value of x, but we are showing that it is being added to 8. If we knew that x were 5, then our sum would be $5 + 8 = 13$. Since $5 + 8 = 8 + 5$, it seems reasonable to conclude that $x + 8 = 8 + x$.

MathFlash!

For any equation dealing with a variable, it is important to remember that although we can assign a value to x, this same value must be used for any other appearance of x. Thus, $x + 8 = 8 + x$ is true because we will only assign the same value of x to each side of the equation. Once you assign 5 to the x on the left side, it must also be assigned to the x on the right side.

2

Example: *How is "the difference between an unknown number and four" expressed in mathematical symbols?*

Solution: Recall that the word "difference" implies subtraction, which is the "−" symbol. The answer is written as $x - 4$. If we knew that x were 7, then the difference would be $7 - 4 = 3$. Note that we could not write $4 - x$, which would have been the correct answer to "the difference between four and an unknown number."

3

Example: *How is "the product of an unknown number and nine" expressed in mathematical symbols?*

Solution: The word "product" is associated with multiplication, so one possible answer would be $9 \times x$. However, in expressions such as this, it is permissible to simply write $9 \cdot x$ or even $9x$. The last of these is the most popular.

4

Example: *How is "the quotient of an unknown number and three" expressed in mathematical symbols?*

Solution: The word "quotient" is associated with division, so the answer is written as $x \div 3$ or as $\frac{x}{3}$. (Either way is acceptable.)

Before proceeding with additional examples, let's list the most common words and/or expressions for the **four basic arithmetic operations**.

1. For **addition**, the most common words and/or expressions are "sum of," "added to," "plus," "more than," and "increased by."

2. For **subtraction**, the most common words and/or expressions are "difference of," "minus," "subtracted from," "subtracted by," "decreased by," and "less than."

3. For **multiplication**, the most common words and/or expressions are "times," "multiplied by," and "product of."

4. For **division**, the most common words and/or expressions are "divided by" and "quotient of."

5

Example: *How is " the product of four and an unknown number subtracted from ten" expressed in mathematical symbols?*

Solution: The phrase "the product of four and an unknown number" translates to $4x$. We know that "subtracted from" deals with the minus symbol. The final answer is $10 - 4x$.

6

Example: *How is "thirteen decreased by the quotient of six and an unknown number" expressed in mathematical symbols?*

Solution: The words "decreased by" mean that we will use the minus sign, with the number 13 appearing before the minus sign. "Quotient" means we must divide, so "quotient of six and an unknown number" becomes $6 \div x \left(\text{or } \dfrac{6}{x} \right)$. The final answer is $13 - \dfrac{6}{x}$ (or $13 - 6 \div x$).

Before we do another example, let's return to Example 6. The final answer was $13 - 6 \div x$. Suppose you knew that the value of x were 3. The statement $13 - 6 \div x$ would then read as $13 - 6 \div 3$. Do you recall how to evaluate such an expression?

Remember that in this instance, division is done before subtraction, so the value of $13 - 6 \div 3$ becomes $13 - 2 = 11$.

7

Example: *How is "twelve increased by the product of an unknown number and two" expressed in mathematical symbols?*

Solution: The words "increased by" means addition, and "the product of an unknown number and two" is translated as $2x$. The final answer is $12 + 2x$.

8

Example: *How is "the quotient of an unknown number and fifteen multiplied by eleven" expressed in mathematical symbols?*

Solution: The first part of this expression would be written as $x \div 15$, and the second part would show a multiplication by the number 11. The final answer would be $x \div 15 \times 11$.

Note that division preceded multiplication.

Example: *How is "sixteen less than three times an unknown number" expressed in mathematical symbols?*

9

Solution: "Three times an unknown number" is translated as $3x$. Since the phrase "sixteen less than" is used, 16 will be subtracted from $3x$, so the final answer will be $3x - 16$. What would the value of this expression be if x were -8? The answer would be $(3)(-8) - 16 = -24 - 16 = -40$.

Example: *How is "twenty times the quotient of seven and an unknown number" expressed in mathematical symbols?*

10

Solution: "Twenty times" becomes $20 \times$. "The quotient of seven and an unknown number" becomes $7 \div x$. Our final answer is $20 \times 7 \div x$ $\left[\text{or } (20)\left(\dfrac{7}{x}\right). \right]$ Do you think that we could also write $140 \times x$? Let's use 5 as a value for x. Sure enough, we find that $20 \times 7 \div 5 = 140 \div 5 = 28$. Thus, $140 \div x$ is a valid answer.

You should know that values of x were assigned here arbitrarily. When you begin studying algebra, you will find that there are situations in which you will find a specific value or set of values for x. Also, when you learn how to solve equations, you will encounter other letters besides x.

 Test Yourself!

Express each of the following using only mathematical symbols. Use x to represent the unknown variable.

1.   The quotient of an unknown number and nineteen

   Answer: _____

2.   Ten increased by an unknown number

   Answer: _____

3.   An unknown number subtracted by five

       Answer: _____

4.   The product of an unknown number and thirteen

       Answer: _____

5.   The difference of six and an unknown number

       Answer: _____

6.   The quotient of two and an unknown number, increased by seven

       Answer: _____

7.   Eleven subtracted from the product of three and an unknown number

       Answer: _____

8.   Four times the quotient of an unknown number and fifteen

       Answer: _____

9.   Twenty more than sixteen times an unknown number

       Answer: _____

10.   One hundred decreased by the quotient of an unknown number
       divided by five

       Answer: _____

QUIZ FIVE

1. How many numbers, between 11 and 40 are divisible by both 2 and 3?

    A 4

    B 5

    C 6

    D 7

2. Which one of the following is <u>not</u> a factor of 150?

    A 30

    B 25

    C 15

    D 12

3. Which one of the following is the correct prime factorization of 4698?

    A $2^2 \times 3^3 \times 29$

    B $2 \times 3^4 \times 29$

    C $2 \times 3^3 \times 29$

    D $2^2 \times 3^4 \times 29$

4. What is the greatest common factor for the numbers 20, 32, and 48?

    A 4

    B 6

    C 8

    D 10

5. What is the least common multiple for the numbers 56 and 294?

    A 588

    B 784

    C 1176

    D 2058

6. Consider the five-digit number 35,67_. Which one of the following digits, when used as the units digit, will create a number that is divisible by 9?

    A 3

    B 4

    C 5

    D 6

7. Which one of the following numbers is divisible by 2, 3, 5, and 7?

    A 420

    B 315

    C 270

    D 168

8. **The number 18 is a factor of which one of the following?**

   A 88

   B 76

   C 60

   D 54

9. **What is the greatest common factor for the numbers 69 and 70?**

   A 1

   B 3

   C 5

   D 9

10. **What is the least common multiple for the numbers 80, 64, and 75?**

    A 2000

    B 3600

    C 4800

    D 8080

CUMULATIVE EXAM

1. What is the decimal form of $5\frac{1}{4}$%?

   A 0.525

   B 0.514

   C 0.0525

   D 0.0514

2. What is the mixed fraction form of $\frac{79}{8}$?

   A $9\frac{1}{8}$

   B $9\frac{3}{8}$

   C $9\frac{7}{8}$

   D $10\frac{1}{8}$

3. Which one of the following has the same value as $\left(-\frac{5}{4}\right)^{12}$?

   A $\left(-\frac{25}{16}\right)^{6}$

   B $\left(-\frac{4}{5}\right)^{12}$

   C $-\frac{60}{48}$

   D $\frac{60}{48}$

4. The current price of a postage stamp is \$0.40. If the price increases by 15% next year, what will be the new price?

   A \$0.42

   B \$0.46

   C \$0.50

   D \$0.55

5. What is the ratio, in reduced form, of 4 days to 100 hours?

   A $\frac{24}{25}$

   B $\frac{23}{24}$

   C $\frac{1}{24}$

   D $\frac{1}{25}$

6. What is the value of $(-3)^0 - 3^2$?

   A −12

   B −9

   C −8

   D −6

7. What is the number 681,529 rounded off to the nearest thousand?

   A 600,000

   B 600,500

   C 681,000

   D 682,000

8. What is the scientific notation for 0.0000085?

   A 85.0 × 10⁻⁷

   B 850 × 10⁻⁸

   C 0.85 × 10⁻⁵

   D 8.5 × 10⁻⁶

9. What is the least common multiple for the numbers 33 and 45?

   A 1485

   B 495

   C 165

   D 3

10. <u>Not</u> including 1 and 42, how many other factors of 42 are there?

    A 8

    B 7

    C 6

    D 5

11. How should the phrase "seven hundred four ten-thousandths" be written in digits?

    A 0.00704

    B 0.0704

    C 0.0740

    D 0.704

12. Sharon had planned to complete $\frac{5}{6}$ of her projects at work today. However, due to other commitments, she only finished $\frac{1}{9}$ of them. What fraction of her projects remain to be completed in order to reach her goal?

    A $\frac{13}{18}$

    B $\frac{7}{9}$

    C $\frac{8}{9}$

    D $\frac{17}{18}$

13. You are given the following list of numbers: $\frac{9}{11}$, $0.8\overline{2}$, $\frac{19}{23}$, and 0.82. Which one of these numbers is largest?

    A $\frac{9}{11}$

    B $0.8\overline{2}$

    C $\frac{19}{23}$

    D 0.82

14. Using x to represent the unknown number, which one of the following is the mathematical description of "thirteen subtracted from the product of nine and an unknown number"?

    A $9x - 13$

    B $9x + 13$

    C $13 - 9x$

    D $13x - 9$

15. Yesterday morning, the temperature was +25°. By that evening, the temperature had dropped by 31°. What was the temperature in the evening?

   A +6°

   B −6°

   C −25°

   D −31°

16. What is the value $4\frac{1}{4} \div \frac{1}{8}$?

   A $\frac{1}{34}$

   B $\frac{17}{32}$

   C $\frac{32}{17}$

   D 34

17. What is the percent equivalent of $\frac{20}{9}$?

   A 22.2%

   B $22.\overline{2}$%

   C 222.2%

   D $222.\overline{2}$%

18. When the number 6615 is written in prime factorization form, what is the exponent for the base 3?

   A 1

   B 2

   C 3

   D 4

19. What is the reduced fraction form of $0.5\frac{2}{5}$?

   A $\frac{27}{50}$

   B $\frac{13}{25}$

   C $\frac{38}{75}$

   D $\frac{9}{20}$

20. Which one of the following numbers, when rounded off to the nearest hundredth, would be written as 0.55?

   A 0.5445

   B 0.5545

   C 0.5554

   D 0.5565

21. Which one of the following statements is completely true?

   A Zero divided by any number is zero.

   B Zero divided by a nonzero number is zero.

   C Zero multiplied by any number is not defined.

   D Zero added to any number is zero.

22. What is the value of $(-5)^2 - (12 - 15 \div 3) \times 3$?

   A 28

   B 22

   C 13

   D 4

23. What is the greatest common factor for the numbers 24, 72, and 40?

    A 4

    B 6

    C 8

    D 12

24. Suppose that a number is divisible by 4. On a sheet of paper, the number appears as 35,5_0. How many different digits could be put in the missing tens place?

    A 5

    B 4

    C 3

    D 2

25. Which one of the following numbers has no other prime factor except 7?

    A 14

    B 21

    C 35

    D 49

Answer Key

1

1. Fifty-eight thousand eight hundred forty-one
2. Nine thousand five
3. Three hundred seven thousandths
4. Six thousand five hundred forty-three ten-thousandths
5. Eight hundred six hundred-thousandths
6. 401,027
7. 1715
8. 0.081
9. 0.00542
10. 0.2409

2

1. 9110 — Since 7 ≥ 5, increase 0 to 1 and change 7 to zero.
2. 2600 — Since 3 < 5, change 3 and 9 to zero.
3. 50,000 — The digit 6 in the hundreds place is larger than 5. Therefore the 9 becomes a 10. This forces the 49 to become 50. Change 661 to all zeros.
4. 5400 — Since 3 < 5, change 3 to zero.
5. 748,500 — Since 9 ≥ 5, increase 4 in the hundreds place to 5 and change 9 and 2 to zero.
6. 905,000 — Since 1 < 5, change 1 and both 8's to zero.
7. 400,000 — Since 9 ≥ 5, increase 3 to 4 and change 9, 6, and all 2's to zero.
8. 15,400 — Look at the tens place. Since 4 < 5, change that 4 and 8 to zero.
9. 16,000 — Look at the hundreds place. Since 0 < 5, change 9 to zero and leave the other zeros alone.
10. 49,900 — Since 2 < 5, change 2 and the units digit 4 to zero.

Lessons

3

1. 3.12   Since 4 < 5, do not change 2. Drop the 4.

2. 0.9   Since 5 ≥ 5, increase 8 to 9. Drop the 5, 3, and 1.

3. 0.7104   Since 8 ≥ 5, increase 3 to 4. Drop the 8.

4. 0.90   Since 6 ≥ 5, increase 9 to 0 and consequently change 8 to 9.
Drop the 6 and 1.

5. 0.429   Since 5 ≥ 5, increase 8 to 9. Drop the 5.

6. 0.62   Since 8 ≥ 5, increase 1 to 2. Drop the 8 and 3.

7. 0.1   Since 9 ≥ 5, increase 0 to 1. Drop the 9, 4, and 5.

8. 0.5455   Look at the hundred-thousandths place. Since 5 ≥ 5, increase 4
(in the ten-thousandths place) to 5. Drop the two rightmost digits.

9. 2.175   Look at the ten-thousandths place. Since 7 ≥ 5, increase 4 to 5.
Drop the two rightmost digits.

10. 0.5   Since 2 < 5, do not change 5. Drop the 2, 9, 8, and 3.

11. 7.508   Since 3 < 5, do not change 8. Drop the 3 and 2.

12. 0.1200   Since 6 ≥ 5, increase the 9 (in the ten-thousandths place) to 0.
This forces the 9 (in the thousandths place) to 0, and thus forces
the 1 (in the hundredths place) to 2.

4

1. −4   Change the problem to (+5) + (−9).

2. −10   Add 6 and 4, retain the minus sign.

3. −6   Take the difference of 9 and 3. Use the minus sign.

4. 8   Take the difference of 15 and 7. Use the plus sign.

5. 7   Change the problem to (−2) + (+9).

6. −21   Change the problem to (−10) + (−11).

7. 29   Write the problem as (+12) − (−17), which becomes (+12) + (+17).

8. −11   Write the problem as (−15) + (+4).

9. 18   Write the problem as (+32) + (−14).

10. 9   Write the problem as (+13) + (+3) + (−7), which becomes (+16) + (−7).

11. −36   Add the values 5, 17, 4, and 10. Retain the minus sign.

12. 23°   Write the problem as (−8) + (+31).

Lessons

5

1. –28   Multiply 7 by 4. (+)(–) = –

2. 55   Multiply 5 by 11. (–)(–) = +

3. –420   Multiply the numbers 3, 14, 2, and 5. (–)(+)(–)(–) = –

4. –6   Divide 90 by 15. (+) ÷ (–) = –

5. 5   Divide 40 by 8. (+) ÷ (+) = +

6. 7   Divide 28 by 4. (–) ÷ (–) = +

7. –6   Divide 96 by 8 to get 12. Then divide 12 by 2. (+) ÷ (–) ÷ (+) = –

8. B   Any number divided by zero is undefined.

9. 0   Any number multiplied by zero results in an answer of zero.

10. $9600   Multiply $800 by 12.

6

1. $\frac{3}{5}$ Divide the numerator and denominator by 4.

2. $\frac{8}{7}$ or $1\frac{1}{7}$ Divide the numerator and denominator by 8.

3. $\frac{5}{13}$ Divide the numerator and denominator by 6.

4. $\frac{3}{23}$ Divide the numerator and denominator by 9.

5. $\frac{15}{11}$ or $1\frac{4}{11}$ Divide the numerator and denominator by 5.

6. $\frac{18}{23}$ Divide the numerator and denominator by 7.

7. $\frac{1}{24}$ Change 10 feet to 120 inches. Reduce $\frac{5}{120}$ by dividing the numerator and denominator by 5.

8. $\frac{64}{7}$ or $9\frac{1}{7}$ Change 4 pounds to 64 ounces.

9. $\frac{3}{20}$ Change 2 hours to 120 minutes. Reduce $\frac{18}{120}$ by dividing the numerator and denominator by 6.

10. $\frac{2}{27}$ Change 9 years to 108 months. Reduce $\frac{8}{108}$ by dividing the numerator and denominator by 4.

11. $\frac{8}{1}$ or 8   Change 2 years to 104 weeks. Reduce $\frac{104}{13}$ by dividing the numerator and denominator by 13.

12. $\frac{5}{9}$ Change 6 yards to 18 feet. Reduce $\frac{10}{18}$ by dividing the numerator and denominator by 2.

Lessons

7

1. $1\frac{8}{9}$ 17 divided by 9 equals 1, with a remainder of 8.

2. $4\frac{1}{3}$ 26 divided by 6 equals 4, with a remainder of 2. Reduce $\frac{2}{6}$ to $\frac{1}{3}$.

3. $8\frac{1}{4}$ 99 divided by 12 equals 8, with a remainder of 3. Reduce $\frac{3}{12}$ to $\frac{1}{4}$.

4. $6\frac{7}{10}$ 134 divided by 20 equals 6, with a remainder of 14. Reduce $\frac{14}{20}$ to $\frac{7}{10}$.

5. $5\frac{4}{17}$ 89 divided by 17 equals 5, with a remainder of 4.

6. $\frac{23}{7}$ Multiply 7 by 3, then add 2 to get the numerator.

7. $\frac{81}{11}$ Multiply 11 by 7, then add 4 to get the numerator.

8. $\frac{49}{10}$ Multiply 10 by 4, then add 9 to get the numerator.

9. $\frac{47}{9}$ Reduce $\frac{4}{18}$ to $\frac{2}{9}$. Multiply 9 by 5, then add 2 to get the numerator.

10. $\frac{44}{5}$ Reduce $\frac{28}{35}$ to $\frac{4}{5}$. Multiply 5 by 8, then add 4 to get the numerator.

11. $\frac{16}{5}$ Reduce $\frac{128}{40}$ by dividing the numerator and denominator by 8.

12. $3\frac{1}{5}$ 16 divided by 5 equals 3, with a remainder of 1.

8

1. $\frac{1}{2}$ — Add the numerators to get $\frac{4}{8}$. Then reduce $\frac{4}{8}$ to $\frac{1}{2}$.

2. $\frac{14}{15}$ — Change the problem to $\frac{5}{15} + \frac{9}{15}$. Add the numerators.

3. $\frac{15}{28}$ — Change the problem to $\frac{36}{28} - \frac{21}{28}$. Subtract the numerators.

4. $\frac{37}{24}$ or $1\frac{13}{24}$ — Change the problem to $\frac{36}{24} + \frac{21}{24} - \frac{20}{24}$. Combine the numerators.

5. $\frac{31}{28}$ or $1\frac{3}{28}$ — Change the problem to $\frac{21}{28} + \frac{10}{28}$. Add the numerators.

6. $\frac{1}{30}$ — Change the problem to $\frac{26}{30} - \frac{25}{30}$. Subtract the numerators.

7. $\frac{113}{36}$ or $3\frac{5}{36}$ — Change the problem to $\frac{84}{36} + \frac{32}{36} - \frac{3}{36}$. Combine the numerators.

8. $6\frac{18}{35}$ or $\frac{228}{35}$ — Change the problem to $3\frac{25}{35} + 2\frac{28}{35}$, which becomes $5\frac{53}{35}$. Convert $\frac{53}{35}$ to $1\frac{18}{35}$ and add to 5.

9. $2\frac{19}{24}$ or $\frac{67}{24}$ — Change the problem to $5\frac{16}{24} - 2\frac{21}{24}$. Rewrite as $4\frac{40}{24} - 2\frac{21}{24}$, then subtract the whole number parts and the numerators of the fraction parts.

10. $7\frac{17}{20}$ or $\frac{157}{20}$ — Change the problem to $6\frac{15}{20} + 2\frac{16}{20} - 1\frac{14}{20}$. Combine the whole numbers to get 7. Combine the numerators to get 17.

11. $\frac{7}{16}$ — Calculate $\frac{3}{4} - \frac{5}{16}$, which is equivalent to $\frac{12}{16} - \frac{5}{16}$.

12. $\frac{43}{45}$ — Calculate $\frac{5}{9} + \frac{2}{5}$, which is equivalent to $\frac{25}{45} + \frac{18}{45}$.

Lessons

9

1. $\frac{3}{16}$ The numerator is (1)(3) and the denominator is (2)(8).

2. $\frac{6}{25}$ Change 4 and 10 to 2 and 5, respectively. The numerator is (2)(3) and the denominator is (5)(5).

3. $\frac{32}{3}$ or $10\frac{2}{3}$ Write the problem as $\frac{16}{7} \times \frac{14}{3}$. Change 7 and 14 to 1 and 2, respectively. The numerator is (16)(2) and the denominator is (1)(3).

4. $\frac{27}{20}$ or $1\frac{7}{20}$ Write the problem as $\frac{9}{10} \times \frac{3}{2}$. The numerator is (9)(3) and the denominator is (10)(2).

5. $\frac{20}{3}$ or $6\frac{2}{3}$ Write the problem as $\frac{8}{5} \times \frac{25}{6}$. Change 8 and 6 to 4 and 3, respectively. Change 5 and 25 to 1 and 5, respectively. The numerator is (4)(5) and the denominator is (1)(3).

6. $\frac{11}{63}$ Write the problem as $\frac{4}{7} \div \frac{36}{11}$, which becomes $\frac{4}{7} \times \frac{11}{36}$. Change 4 and 36 to 1 and 9, respectively. The numerator is (1)(11) and the denominator is (7)(9).

7. 36   Write the problem as $\frac{4}{5} \times \frac{6}{1} \times \frac{15}{2}$. Change 4 and 2 to 2 and 1, respectively. Change 5 and 15 to 1 and 3, respectively. The numerator is (2)(6)(3) and the denominator is (1)(1)(1).

8. $\frac{11}{9}$ or $1\frac{2}{9}$ Change the problem to $\frac{11}{8} \times \frac{4}{1} \times \frac{2}{9}$. Change each of the 4 , 2, and 8 to 1. The numerator is (11)(1)(1) and the denominator is (1)(1)(9).

9. $\frac{16}{3}$ or $5\frac{1}{3}$ acres of land   Calculate $17\frac{1}{3} \times \frac{4}{13}$, which is equivalent to $\frac{52}{3} \times \frac{4}{13}$. Change 52 and 13 to 4 and 1, respectively. The numerator is (4)(4) and the denominator is (3)(1).

10. $\frac{36}{5}$ or $7\frac{1}{5}$ miles per hour   Calculate $28\frac{4}{5} \div 4$, which is equivalent to $\frac{144}{5} \times \frac{1}{4}$. Change 144 and 4 to 36 and 1, respectively. The numerator is (36)(1) and the denominator is (5)(1).

Lessons

10

1. 0.02   Move the decimal point two places to the left.

2. 0.194   Move the decimal point two places to the left.

3. 0.00059   Move the decimal point two places to the left.

4. 2.05   Move the decimal point two places to the left.

5. 0.12125   Change $\frac{1}{8}$ to .125 so that the problem reads 12.125%.
Move the decimal point two places to the left.

6. 50.01   Move the decimal point two places to the left.

7. $\frac{4}{25}$ Write the problem as $\frac{16}{100}$, then reduce to lowest terms by dividing the numerator and denominator by 4.

8. $1\frac{2}{5}$ or $\frac{7}{5}$ Write the problem as $\frac{140}{100}$, then reduce to lowest terms by dividing the numerator and denominator by 20.

9. $\frac{1}{25,000}$ Write the problem as $\frac{.004}{100} = \frac{4}{100,000}$, then reduce to lowest terms by dividing the numerator and denominator by 4.

10. $\frac{147}{2000}$ Write the problem as $\frac{7.35}{100} = \frac{735}{10,000}$, then reduce to lowest terms by dividing the numerator and denominator by 5.

11. $\frac{121}{500}$ Write the problem as $\frac{24.2}{100} = \frac{242}{1000}$, then reduce to lowest terms by dividing the numerator and denominator by 2.

12. $\frac{1}{550}$ Write the problem as $\frac{\frac{2}{11}}{100} = \frac{2}{11} \times \frac{1}{100} = \frac{2}{1100}$, then reduce to lowest terms by dividing the numerator and denominator by 2.

Lessons

11

1. 27.5%   Move the decimal point two places to the right.

2. 870%   Move the decimal point two places to the right.

3. 4321%   Move the decimal point two places to the right.

4. 0.0567%   Move the decimal point two places to the right.

5. 4.125%   Change $\frac{1}{8}$ to .125 so that the problem is 0.04125. Move the decimal point two places to the right.

6. $\frac{1}{125}$ Write the problem as $\frac{8}{1000}$, then reduce to lowest terms by dividing the numerator and denominator by 8.

7. $\frac{13}{5000}$ Write the problem as $\frac{26}{10,000}$, then reduce to lowest terms by dividing the numerator and denominator by 2.

8. $1\frac{12}{25}$ or $\frac{37}{25}$ Write the problem as $1\frac{48}{100}$, then reduce $\frac{48}{100}$ to lowest terms by dividing the numerator and denominator by 4.

9. $\frac{19}{8000}$ Write the problem as $\frac{2\frac{3}{8}}{1000} = \frac{\frac{19}{8}}{1000}$, which becomes $\frac{19}{8} \times \frac{1}{1000} = \frac{19}{8000}$.

10. $\frac{121}{12,000}$ Write the problem as $\frac{10\frac{1}{12}}{1000} = \frac{\frac{121}{12}}{1000}$, which becomes $\frac{121}{12} \times \frac{1}{1000} = \frac{121}{12,000}$

Lessons

12

1. $0.\overline{63}$ Divide 11 into 7.00

2. 1.625   Divide 8 into 13.00

3. $0.41\overline{6}$ Divide 12 into 5.00

4. 0.00375   Divide 800 into 3.00

5. $0.\overline{054}$ Divide 111 into 6.00

6. 220%   Divide 5 into 11 to get 2.2
Move the decimal point two places to the right.

7. 56.25%   Divide 16 into 9 to get 0.5625
Move the decimal point two places to the right.

8. $0.\overline{75}\%$ Divide 3300 into 25 to get $0.00\overline{75}$
Move the decimal point two places to the right.

9. $18.\overline{8}\%$ Divide 90 into 17 to get $0.18\overline{8}$
Move the decimal point two places to the right.

10. $393.\overline{3}\%$ Divide 15 into 59 to get $3.9\overline{3}$
Move the decimal point two places to the right.

13

1. 110   Multiply 88 by 1.25
An increase of 25% is equivalent to a multiplication of 125%.

2. 191.1   Multiply 175 by 1.092
An increase of 9.2% is equivalent to a multiplication of 109.2%.

3. 11.1672   Multiply 10.8 by 1.034
An increase of 3.4% is equivalent to a multiplication of 103.4%.

4. 16.64   Multiply 6.4 by 2.60
An increase of 160% is equivalent to a multiplication of 260%.

5. 29.16   Multiply 36 by 0.81
A decrease by 19% is equivalent to a multiplication of 100% − 19% = 81%.

6. 78   Multiply 260 by 0.30
A decrease by 70% is equivalent to a multiplication of 100% − 70% = 30%.

7. 900.882   Multiply 904.5 by 0.996
A decrease by 0.4% is equivalent to a multiplication of 100% − 0.4% = 99.6%

8. 48.204   Multiply 52 by 0.927
A decrease by 7.3% is equivalent a multiplication of 100% − 7.3% = 92.7%.

9. 189 books   Multiply 140 by 1.35
An increase of 35% is equivalent to a multiplication of 135%.

10. $96.46   Multiply $106 by 0.91
A discount of 9% is equivalent to a multiplication of 100% − 9% = 91%.

Lessons

14

1. 0.0885, 0.09, 0.7889, 0.7989

   Write each decimal in ten-thousandths. 0.09 is equivalent to 0.0900

2. 0.07, 0.108, 0.56, 0.6

   Write each decimal in thousandths. 0.07 becomes 0.070, 0.56 becomes 0.560, and 0.6 becomes 0.600

3. 0.1452, 0.1524, 0.251, 0.3, 0.42

   Write each decimal in ten-thousandths. 0.251 becomes 0.2510, 0.3 becomes 0.3000, and 0.42 becomes 0.4200

4. $\frac{5}{12}, \frac{7}{16}, \frac{16}{25}, \frac{2}{3}$

   Change each fraction to its decimal equivalent.
   $\frac{5}{12} = 0.41\overline{6}$, $\frac{7}{16} = 0.4375$, $\frac{16}{25} = 0.64$, and $\frac{2}{3} = 0.\overline{6}$

5. $\frac{10}{23}, \frac{4}{9}, \frac{16}{33}, \frac{1}{2}, \frac{13}{24}$

   Change each fraction to its decimal equivalent.
   $\frac{10}{23} \approx 0.4345$, $\frac{4}{9} = 0.\overline{4}$, $\frac{16}{33} = 0.\overline{48}$, $\frac{1}{2} = 0.5$, and $\frac{13}{24} = 0.541\overline{6}$

6. $0.6, \frac{11}{18}, \frac{5}{7}, 0.72, \frac{3}{4}$

   Change the fractions to their decimal equivalents.
   $\frac{11}{18} = 0.6\overline{1}$, $\frac{5}{7} \approx 0.7143$, and $\frac{3}{4} = 0.75$

7. $\frac{3}{28}, \frac{1}{9}, \frac{2}{17},$ 0.127, 0.1288

   Change the fractions to their decimal equivalents.
   $\frac{3}{28} \approx 0.1071$, $\frac{1}{9} = 0.\overline{1}$, and $\frac{2}{17} \approx 0.1176$

8. $0.0954, \frac{4}{27}, 0.149,$ $0.154, \frac{39}{250}$

   Change the fractions to their decimal equivalents.
   $\frac{4}{27} = 0.\overline{148}$ and $\frac{39}{250} = 0.156$

9. Any one of the numbers 0.979, 0.980, or 0.981

   0.9_ _ The only digits that fit are 79, 80, or 81.

10. Either of the numbers 0.841 or 0.842

    0.84_ The only digits that fit are 1 or 2.

Lessons

15

1. $\left(-\dfrac{8}{3}\right)^3$ The fraction is multiplied by itself 3 times.

2. $\dfrac{256}{81}$ or $3\dfrac{13}{81}$ $\left(-\dfrac{4}{3}\right)\left(-\dfrac{4}{3}\right)\left(-\dfrac{4}{3}\right)\left(-\dfrac{4}{3}\right)$

3. $1,419,857$ $(17)(17)(17)(17)(17)$

4. $-59,049$ $(-9)(-9)(-9)(-9)(-9)$

5. 169 $(-13)(-13)$

6. $-\dfrac{1}{2,097,152}$ $\left(-\dfrac{1}{8}\right)\left(-\dfrac{1}{8}\right)\left(-\dfrac{1}{8}\right)\left(-\dfrac{1}{8}\right)\left(-\dfrac{1}{8}\right)\left(-\dfrac{1}{8}\right)\left(-\dfrac{1}{8}\right)$

7. $\dfrac{25}{16}$ or $1\dfrac{9}{16}$ $\left(-\dfrac{5}{4}\right)\left(-\dfrac{5}{4}\right)$

8. $\dfrac{1331}{27}$ or $49\dfrac{8}{27}$ $\left(\dfrac{11}{3}\right)\left(\dfrac{11}{3}\right)\left(\dfrac{11}{3}\right)$

9. $16,777,216$ $(16)(16)(16)(16)(16)(16)$

10. $\dfrac{2401}{16}$ or $150\dfrac{1}{16}$ $\left(\dfrac{7}{2}\right)\left(\dfrac{7}{2}\right)\left(\dfrac{7}{2}\right)\left(\dfrac{7}{2}\right)$

16

1. 1.9×10^1 $1.9 \times 10^1 = (1.9)(10)$
2. 2.6×10^{-2} $2.6 \times 10^{-2} = 2.6 \div 100$
3. 3.457×10^4 $3.457 \times 10^4 = (3.457)(10,000)$
4. 8.43×10^{-4} $8.43 \times 10^{-4} = 8.43 \div 10,000$
5. 9.005111×10^6 $9.005111 \times 10^6 = (9.005111)(1,000,000)$
6. 1.35 Since $10^0 = 1$, $1.35 \times 10^0 = (1.35)(1)$
7. 0.00000004021 $4.021 \times 10^{-8} = 4.021 \div 100,000,000$
8. $68,000,000$ $6.8 \times 10^7 = (6.8)(10,000,000)$
9. 0.00001888 $1.888 \times 10^{-5} = 1.888 \div 100,000$
10. $30,000,000,000$ $3.0 \times 10^{10} = (3.0)(10,000,000,000)$

Lessons

17

1. 22 $12 \div 3 + 3 \times 6 = 4 + 3 \times 6 = 4 + 18$

2. 42 $18 + 4^3 - 8 \times 5 = 18 + 64 - 8 \times 5 = 18 + 64 - 40$

3. 29 $9 - 4 \times (-5) = 9 + 20$

4. −7 $(60 - 5^2) \div 7 - 12 = (60 - 25) \div 7 - 12 = 35 \div 7 - 12 = 5 - 12$

5. −120 $90 \div 6 - 3^3 \times 5 = 90 \div 6 - 27 \times 5 = 15 - 27 \times 5 = 15 - 135$

6. 860 $20 \times (40 - 10 \div 2 + 8) = 20 \times (40 - 5 + 8) = 20 \times 43$

7. −11 $(-1)^8 + (6 - 12 \div 4 \times 6) = 1 + (6 - 12 \div 4 \times 6) = 1 + (6 - 3 \times 6)$
$= 1 + 6 - 18$

8. −150 $-15 - 5 \times 3^3 = -15 - 5 \times 27 = -15 - 135$

9. −66 $28 \div (4 - 18) - 2^6 = 28 \div (-14) - 2^6 = 28 \div (-14) - 64 = -2 - 64$

10. 17 $(-4)^3 + (-18 \div 3 + 3 \times 5)^2 = -64 + (-18 \div 3 + 3 \times 5)^2$
$= -64 + (-6 + 3 \times 5)^2 = -64 + (-6 + 15)^2 = -64 + 9^2 = -64 + 81$

18

1. B   The number named by the last two digits, 32, is divisible by 4.

2. D   Since the number must end in 0 or 5, it must also be divisible by 5.

3. D   The sum of the digits is 21, which is divisible by 3 but is not divisible by 9.

4. C   When the blank is replaced by either 0 or 8, the number named by the last three digits becomes either 360 or 368. Each of these numbers is divisible by 8.

5. C   In order for a number to be divisible by 6, it must be divisible by 2 and by 3.

6. A   To be divisible by 5, the units digit must be 0 or 5. If the digit is 5, the number cannot be even.

7. B   The number 5292 is divisible by 6. Each of the numbers 5090, 5494, and 5696 is not divisible by 6.

8. A   30 is divisible by each of 2, 3, and 5.

9. C   Each of the three numbers 6711, 6741, and 6771 is divisible by 3.

10. B   60 is divisible by both 3 and 4, but is not divisible by 9.

Lessons

19

1.	D	The primes are 2, 3, 5, and 7.
2.	29	The numbers 27 and 28 are composite.
3.	13, 23, 43, 53, 73, 83	The numbers 33, 63, and 93 are composite.
4.	1, 3, 13, 39	These are the four factors of 39.
5.	1, 2, 3, 6, 9, 18	These are the six factors of 18.
6.	C	26 is not a factor of 60.
7.	15, 25, 75	The other factors of 75, namely 1, 3, and 5 are not composite.
8.	C	The three composite numbers are 27, 57, and 77.
9.	2	The number 2 is the only even prime.
10.	2, 13	The other factors of 52, namely 1, 4, 26, and 52 are not prime.

20

1.	$2^3 \times 5$	One possible solution: $40 = (10)(4) = (2 \times 5)(2^2)$
2.	$2^3 \times 3 \times 7$	One possible solution: $168 = (8)(21) = (2^3)(3 \times 7)$
3.	$3^2 \times 7$	One possible solution: $63 = (9)(7) = (3^2)(7)$
4.	$5^2 \times 13$	One possible solution: $325 = (5)(65) = (5)(5 \times 13)$
5.	7×13	The only solution: $91 = (7)(13)$
6.	$7^2 \times 11$	One possible solution: $539 = (11)(49) = (11)(7^2)$
7.	$2^4 \times 11 \times 17$	One possible solution: $2992 = (8)(374) = (8)(11)(34) = (2^3)(11)(2)(17)$
8.	$3^3 \times 23$	One possible solution: $621 = (9)(69) = (3^2)(3)(23)$
9.	$2^5 \times 7$	One possible solution: $224 = (8)(28) = (2^3)(2^2)(7)$
10.	$2 \times 19 \times 31$	One possible solution: $1178 = (2)(589) = (2)(19)(31)$

Lessons

21

1. 6 $30 = 2 \times 3 \times 5$ and $42 = 2 \times 3 \times 7$
Use 2×3

2. 10 $160 = 2^5 \times 5$ and $90 = 2 \times 3^2 \times 5$
Use 2×5

3. 27 $135 = 3^3 \times 5$ and $297 = 3^3 \times 11$
Use 3^3

4. 13 $91 = 7 \times 13$ and $156 = 2^2 \times 3 \times 13$
Use 13

5. 1 $25 = 5^2$ and $51 = 3 \times 17$
Use 1

6. 15 $120 = 2^3 \times 3 \times 5$ and $345 = 3 \times 5 \times 23$
Use 3×5

7. 41 $41 =$ Prime and $656 = 2^4 \times 41$
Use 41

8. 40 $280 = 2^3 \times 5 \times 7$ and $2080 = 2^5 \times 5 \times 13$
Use $2^3 \times 5$

9. 5 $35 = 5 \times 7$, $65 = 5 \times 13$, and $60 = 2^2 \times 3 \times 5$
Use 5

10. 54 $270 = 2 \times 3^3 \times 5$, $216 = 2^3 \times 3^3$, and $486 = 2 \times 3^5$
Use 2×3^3

22

1. 210 $30 = 2 \times 3 \times 5$ and $42 = 2 \times 3 \times 7$
Use $2 \times 3 \times 5 \times 7$

2. 1440 $160 = 2^5 \times 5$ and $90 = 2 \times 3^2 \times 5$
Use $2^5 \times 3^2 \times 5$

3. 1485 $135 = 3^3 \times 5$ and $297 = 3^3 \times 11$
Use $3^3 \times 5 \times 11$

4. 1092 $91 = 7 \times 13$ and $156 = 2^2 \times 3 \times 13$
Use $2^2 \times 3 \times 7 \times 13$

5. 1275 $25 = 5^2$ and $51 = 3 \times 17$
Use $3 \times 5^2 \times 17$

6. 2760 $120 = 2^3 \times 3 \times 5$ and $345 = 3 \times 5 \times 23$
Use $2^3 \times 3 \times 5 \times 23$

7. 656 $41 =$ Prime and $656 = 2^4 \times 41$
Use $2^4 \times 41$

8. 14,560 $280 = 2^3 \times 5 \times 7$ and $2080 = 2^5 \times 5 \times 13$
Use $2^5 \times 5 \times 7 \times 13$

9. 5460 $35 = 5 \times 7$, $65 = 5 \times 13$, and $60 = 2^2 \times 3 \times 5$
Use $2^2 \times 3 \times 5 \times 7 \times 13$

10. 9720 $270 = 2 \times 3^3 \times 5$, $216 = 2^3 \times 3^3$, and $486 = 2 \times 3^5$
Use $2^3 \times 3^5 \times 5$

Lessons

23

1. $\frac{x}{19}$ "Quotient of" means "division."

2. $10 + x$ or $x + 10$ "Increased by" means "added to."

3. $x - 5$ "Subtracted by" means "minus."

4. $13x$ "Product of" means "multiplication."

5. $6 - x$ "Difference of" means "subtraction."

6. $\frac{2}{x} + 7$ or $7 + \frac{2}{x}$ "Quotient of" means "division" and "increased by" means "added to."

7. $3x - 11$ "Subtracted from" means "minus," but reverse the order of the numbers. "Product of" means "multiplication."

8. $4 \times \frac{x}{15}$ or $\frac{4x}{15}$ "Times" means "multiplication" and "quotient of" means "division."

9. $20 + 16x$ or $16x + 20$ "More than" means "added to" and "times" means "multiplication."

10. $100 - \frac{x}{5}$ "Decreased by" means "minus."

Quizzes

1

1. C Four decimal places means "ten-thousandths."

2. A "Seven hundred two thousand" means 702,000. Just add 60.

3. B Write the problem as (–25) + (+11).

4. D 750 – (–282).

5. B Split the problem into 40,000 and 809.

6. B Split the problem into "two hundred thousandths" and "six thousandths."

7. C (–6 + 11) – (+17) = (5) + (–17).

8. D The result of any number divided by zero is undefined.

9. A Since 9 >5, increase 8 to 9.

10. A Since 4 < 5, do not change 6. Change the rightmost three digits to zeros.

2

1. D Change 7 yards to 21 feet.

2. B 91 ÷ 5 = 18, with a remainder of 1. The remainder is placed over 5.

3. D Change the problem to $\frac{16}{44} + \frac{33}{44}$.

4. C Other than 1, there is no common factor for 35 and 39.

5. C Change the problem to

$$7\frac{21}{28} = \overset{6}{\cancel{7}}\frac{21}{28} + \frac{28}{28} = 6\frac{49}{28}$$
$$-2\frac{24}{28} \qquad -2\frac{24}{28} \qquad -2\frac{24}{28}$$

6. A Change the problem to $\frac{80}{12} - \frac{21}{12} + \frac{30}{12}$.

7. D Change 10 and 15 to 2 and 3. The problem becomes $\frac{7}{3} \times \frac{2}{13}$.

8. B Change the problem to $\frac{18}{5} \times \frac{5}{24}$, which simplifies to $\frac{3}{1} \times \frac{1}{4}$.

9. A Change the problem to $\frac{9}{5} \times \frac{3}{1} \times \frac{8}{33}$, which simplifies to $\frac{9}{5} \times \frac{1}{1} \times \frac{8}{11}$.

10. D $\frac{5}{6} \times \frac{1}{8}$, since "of" means "times."

3

1. C $\dfrac{.018}{100} = \dfrac{18}{100,000}$. Reduce to lowest terms.

2. A $8\dfrac{1}{5}\% = 8.2\%$. Move the decimal point two places to the left.

3. A $\dfrac{27}{16} = 1.6875$. Move the decimal point two places to the right.

4. B   6 divided by 125 yields 0.048

5. C $0.361 < 0.371 < 0.3721$

6. D $\dfrac{17}{37} < 0.469 < 0.47 < \dfrac{10}{21} < \dfrac{13}{27}$

7. C $0.07\dfrac{2}{3} = \dfrac{7\dfrac{2}{3}}{100} = \dfrac{23}{3} \times \dfrac{1}{100}$

8. B   (48)(0.78). A decrease by 22% means a multiplication of $100\% - 22\% = 78\%$.

9. B   (2000)(1.0075). An increase of 0.75% means a multiplication of 100.75%.

10. D   ($9.00)(1.144). An increase of 14.4% means a multiplication of 114.4%. Round off to the nearest hundredth.

4

1. A $1^7 - (-1)^5 = 1 - (-1)$

2. B $(4.83)(10^{-4}) = 4.83 \div 10,000$

3. D   Insert the decimal point between 5 and 2. The exponent on the base 10 is the number of places to the right to reach 526,000.

4. B $(85 - 81) + 20 \div (-2) = 4 + (-10)$

5. D $-8 - (+36) \div 4 + (-1) = -8 - (+9) + (-1)$

6. C   Insert the decimal point between 1 and 0. The exponent on the base 10 is the number of places to the right to reach 10.

7. C   As an integer, the decimal point must be moved 15 places to the right. The number would be 4007, followed by 12 zeros.

8. A $\left(-\dfrac{13}{3}\right)\left(-\dfrac{13}{3}\right)$

9. C $(-2)^3 - \left(\dfrac{1}{2}\right)^2 = -8 - \dfrac{1}{4}$

10. B   Use the answer choice with the smallest exponent for the base 10. Note that $-10 < -9 < -8 < -7$.

Quizzes

5

1. B   The five numbers are 12, 18, 24, 30, and 36.

2. D   150 does not divide evenly by 12.

3. B $2 \times 3^4 \times 29 = 2 \times 81 \times 29 = 4698$

4. A   In factored form, the numbers are $2^2 \times 5$, 2^5, and $2^4 \times 3$.
Use 2^2 for the GCF.

5. C   In factored form, the numbers are $2^3 \times 7$ and $2 \times 3 \times 7^2$.
Use $2^3 \times 3 \times 7^2$ for the LCM.

6. D   The sum of the digits of 35,676 is 27, and 27 is divisible by 9.

7. A $420 = 2^2 \times 3 \times 5 \times 7$, so it is divisible by each of 2, 3, 5, and 7.

8. D   54 divides evenly by 18.

9. A $69 = 3 \times 23$ and $70 = 2 \times 5 \times 7$. There are no common bases,
so the GCF is 1.

10. C   In factored form, the numbers are $2^4 \times 5$, 2^6, and 3×5^2.
Use $2^6 \times 3 \times 5^2$ for the LCM.

Cumulative Exam

1. **C**   Change to 5.25%. Move the decimal point two places to the left. (Percent to decimal)

2. **C**   79 divided by 8 yields 9 with a remainder of 7. (Improper to mixed fraction)

3. **A** $\left(-\dfrac{25}{16}\right)^6 = \left(-\dfrac{5^2}{4^2}\right)^6 = \left(-\dfrac{5}{4}\right)^{12}$ (Exponents)

4. **B**   ($0.40)(1.15) (Percent increase)

5. **A**   4 days = (4)(24) = 96 hours. Reduce $\dfrac{96}{100}$ to lowest terms. (Reduction of fractions)

6. **C** $(-3)^0 - 3^2 = 1 - 9$ (Exponents)

7. **D**   Since $5 \geq 5$, increase 1 to 2. Change the three rightmost digits to zeros. (Rounding off integers)

8. **D**   Place the decimal point between 8 and 5. The exponent on 10 will be the negative of the number of places to the left to reach 0.0000085. (Scientific notation)

9. **B** $33 = 3 \times 11$ and $45 = 3^2 \times 5$. Use $3^2 \times 5 \times 11$ as the LCM (Least Common Multiple)

10. **C**   The factors are 2, 3, 6, 7, 14, and 21. (Primes and composites)

11. **B**   "Ten-thousandths" means four decimal places. (Place value for decimals)

12. **A** $\dfrac{5}{6} - \dfrac{1}{9}$. (Addition, subtraction of fractions)

13. **C** $\dfrac{9}{11} < 0.82 < 0.8\overline{2} < \dfrac{19}{23}$
    (Comparison of sizes of fractions and decimals)

Cumulative Exam

14. A "Subtracted from" means "minus," but reverse the order. "Product of" means "multiplication." (Mathematical expressions from words)

15. B (+25) – (31) (Addition, subtraction of signed numbers)

16. D Write the problem as $\dfrac{17}{4} \times \dfrac{8}{1}$ (Multiplication, division of fractions)

17. D Divide 20 by 9 to get $2.\overline{2}$. Move the decimal point two places to the right. (Fraction to percent)

18. C $6615 = 3^3 \times 5 \times 7^2$ (Prime factorization)

19. A Write the problem as 0.54, which becomes $\dfrac{54}{100}$. Reduce to lowest terms. (Decimal to fraction)

20. B Since 4 < 5, drop the two rightmost digits. (Rounding off decimals)

21. B $0 \div x$, where $x \neq 0$, is always 0. (Multiplication, division of signed numbers)

22. D $(-5)^2 - (12 - 15 \div 3) \times 3 = 25 - (12 - 5) \times 3 = 25 - 7 \times 3$ (Order of operations)

23. C The factored form of the numbers are $2^3 \times 3$, $2^3 \times 3^2$, and $2^3 \times 5$ Use 2^3 as the GCF. (Greatest common factor)

24. A The five possible digits are 0, 2, 4, 6, and 8. (Divisibility rules for integers)

25. D $49 = 7^2$, so 7 is the only prime factor of 49. (Primes and composites)

Workspace

Workspace

Workspace

Workspace

Workspace

Workspace

Workspace

Workspace

Workspace

Workspace

Workspace

Workspace

Workspace

SCORECARD
Numbers & Operations

Lesson	Completed	Number of Drill Questions	Number Correct	What I need to review...
1		10		
2		10		
3		12		
4		12		
5		10		
6		12		
7		12		
8		12		
9		10		
10		12		
11		10		
12		10		
13		10		
14		10		
15		10		
16		10		
17		10		
18		10		
19		10		
20		10		
21		10		
22		10		
23		10		

Quiz		What I need to review...
1	/10	
2	/10	
3	/10	
4	/10	
5	/10	

Cumulative Exam	/25	